INJURY AND SKELETAL BIOMECHANICS

Edited by **Tarun Goswami**

Injury and Skeletal Biomechanics
http://dx.doi.org/10.5772/2766
Edited by Tarun Goswami

Contributors

Karel Jelen, František Lopot, Daniel Hadraba, Hynek Herman, Martina Lopotova, Tadayoshi Aoyama, Taisuke Kobayashi, Zhiguo Lu, Kosuke Sekiyama, Yasuhisa Hasegawa, Toshio Fukuda, Andrzej Mroczkowski, Yuri Moskalenko, Gustav Weinstein, Tamara Kravchenko, Peter Halvorson, Natalia Ryabchikova, Julia Andreeva, Orlin Filipov, Aalap Patel, Tarun Goswami, Mary E. Blackmore, Tarun Goswami, Carol Chancey, Maxime Raison, Maria Laitenberger, Aurelie Sarcher, Christine Detrembleur, Jean-Claude Samin, Paul Fisette, J. Mizrahi, D. Daily, Tomáš Gregor, Petra Kochová, Lada Eberlová, Lukáš Nedorost, Eva Prosecká, Václav Liška, Hynek Mírka, David Kachlík, Ivan Pirner, Petr Zimmermann, Anna Králíčková, Milena Králíčková, Zbyněk Tonar, Nancy S. Landínez-Parra, Diego A. Garzón-Alvarado, Juan Carlos Vanegas-Acosta

Published by InTech
Janeza Trdine 9, 51000 Rijeka, Croatia

Publishing Process Manager Marijan Polic
Technical Editor InTech Prepress, Novi Sad
Cover InTech Design team

Additional hard copies can be obtained from orders@intechopen.com

Injury and Skeletal Biomechanics, Edited by Tarun Goswami
p. cm.
ISBN 978-953-51-0690-6

For EU product safety concerns:
IN TECH d.o.o., Prolaz Marije Krucifikse Kozulić 3, 51000 Rijeka, Croatia,
info@intechopen.com or visit our website at intechopen.com.

Meet the editor

Dr Tarun Goswami is a professor of Biomedical, Industrial and Human Factors Engineering, and Orthopaedic Surgery, Sports Medicine, and Rehabilitation at Wright State University, Dayton, OH. Prior to joining WSU, he was an Associate Professor and Chairman of Mechanical Engineering Department, at Ohio Northern University, Ada, OH. Dr. Goswami has diverse research interests in orthopaedic biomechanics, biomaterials, medical devices, and developing new bio-markers to predict various medical conditions. He has active grants from U.S. Air Force and Army. He has published more than 80 articles in archival journals and over 100 presentations, posters, and articles in the proceedings. All his degrees are mechanical engineering, earning a D.Sc. from Helsinki University of Technology.

Contents

Preface

The field of biomechanics has been evolving from the times of ancient Greeks. Recent publications and research in biomechanics sky rocketed as the field of traditional biomechanics is creating new opportunities in diagnostics, therapy, rehabilitation, motion preservation, kinesiology, total joint replacement, biomechanics of living systems at small scale, and other areas.

Biomechanics now encompasses a range of fields. The book on Injury and Skeletal Biomechanics is a broad topic and may provide a platform for newer texts and editions as the research evolves and new results are obtained. In the current form, the book covers four areas: 1) Motion Preservation, which will be useful in designing rehabilitation and training segments, 2) Musculoskeletal and Injury Biomechanics, which includes spine and brain, their behavior under the actions of force, motion, strain, and modeling them analytically and experimentally, 3) Gait-Behavior, is another area which is being developed to learn more on kinesiology and movements of the body, and 4) Quantitative Biomechanics, a somewhat new area that uses imaging and analytical computational tools.

Therefore, the book presents information in four sections, in a concise format. Based on these sections, new courses may be developed at graduate level or some of the concepts used to teach undergraduate students in biomedical engineering. Since the book will be available under open access model, its use will be free to students, and this topic may be introduced as a new course, if desired. The four sections presented in this book will continue to challenge both the researchers and students in the future and therefore, create new knowledge.

Tarun Goswami
Spine Research Group
Biomedical, Industrial and Human Factors Engineering Department
Wright State University
USA

Section 1

Motion Preservation

Chapter 1

The Women's Pelvic Floor Biomechanics

Karel Jelen, František Lopot, Daniel Hadraba,
Hynek Herman and Martina Lopotova

Additional information is available at the end of the chapter

http://dx.doi.org/10.5772/47773

1. Introduction

The function of the pelvic floor is fundamentally influenced by the behaviour of several organs and the organ-linked processes. The aim of this work is to study the properties and changes of the women's pelvic floor. The motive arises from the fact that pelvic floor dysfunctions badly influence the quality of life. The loss of the proper function in the pelvic floor results in a wide range of problems from asymptomatic and anatomic defects to vaginal eversion. All the aforementioned problems are frequently followed by urinating and defecating difficulties together with sexual dysfunctions.

As the initial symptoms of pelvic floor dysfunctions are very weak, the absence of seeking medical assistance among women is significant at the beginning. However, the fact is that an early and explicit diagnosis is crucial. For example, the prevalence of uterovaginal prolapse is about 50 % among delivering women, but only one half of them search for medical care. These types of health problems occur more frequently as the population is aging.

The basis and origins of pelvic floor dysfunctions have certainly a multifactorial character. The elementary factor is intra-abdominal pressure dynamics and it is usually highlighted by obesity, chronic constipation, physically hard work, coughs and mainly pregnancy, vaginal delivery respectively. The topical application of mechanical stress affects the tissue essentially and can make progress towards the failure of tissue continuity. The only solution is usually surgery that tries to fix found problems, revive functional supports of organs and restore their physiological features. From this point of view, the most important area for research on the pelvic floor is the interaction between individual organs (endopelvic fascia mainly) and rheological description of these interactions.

2. Context and paper targets

In pregnancy, a large number of changes are observed in the female body. The main reason for the changes is to cope with the growing foetus's demands and also to protect the woman's health. The changes are mostly controlled by the endocrine system (hypnosis, adrenal, and thyroid glands, placenta, etc.). The system modifies the production of hormones which influence the whole body. The hormonal activity changes the mechanical properties of tissues and together with anatomical modification (growing) affect body posture. One of the organs that are directly impacted is the pelvic floor.

The women's pelvic floor is traditionally defined as a ligament-muscular apparatus that provides a dynamic support to the urethra, bladder, vagina and rectum. It can be divided into the supporting and suspensory parts.

The supporting part is formed by muscles (m. coccygeus, m. levator ani) that create a thin funnel. The funnel is ended by a hole which establishes a corridor for above mention organs. M. levator ani is directly connected to the vaginal muscle. According to the phylogenetic view, the coccygeus muscle (m. coccygeus) is a skeletal muscle and therefore it is directly connected to the musculoskeletal system.

The suspensory part is a fibrous component that is termed the endopelvic fascia. It is a coherent system that surrounds the vagina and connects to the pelvic walls. The fascia's segments are conservatively named the pubocervical fascia, rectovaginal fascia, cardinal ligaments, and sacrouterine ligaments.

The aforementioned muscles and ligaments guarantee the proper function of the pelvic floor. When the function is unbalanced, it causes a fall and disorganization of organs. These changes strongly affect body posture. While muscular problems are usually solved by suitable physiotherapy treatment, problems of the suspensory apparatus are mostly fixed by surgical approaches when an implant is frequently installed.

This paper discusses the influence of pregnancy on the pelvic floor. In and after pregnancy the pelvic floor is even more loaded and stressed and therefore the eventual dysfunctions multiply related unpleasant effects. The main goal is to discover the structural disorders of suspensory apparatus and rheological expression of endopelvic fascia properties. The outcome of this study helps to design better implants and which mechanical properties are not dangerous due to increasing local mechanical stress.

3. Research

The research in this area has been supported by several grants and it is widely discussed in doctoral and master theses within the department. The experiments are measured in a laboratory that is fully equipped for kinematic and dynamic testing as well as for identifying the rheological properties of soft tissues.

3.1. Changes in body posture

The changes in body posture are observed while walking, standing or performing specific movements (for example landing on the heels after standing on tiptoe). The experiments are conducted on women at different stages of pregnancy. This is very important due to the hormonal changes.

In pregnancy the whole musculoskeletal system is influenced by relaxin, which is produced by the placenta, and corpus luteum. They both control the ligamentary apparatus by inhibiting collagen synthesis that amplifies the activity of collagenase and consequently the ligaments of the pelvic girdle and spine become looser. The loose ligaments and weight of the pregnant uterus increase lumbar lordosis. The whole process results in modifications of movement stereotypes. The modification does not only arise from mechanical principles but in particular form the urgency of seeking a relieving posture. A significant role also played by the fact that m. levatoru ani and the thoracic muscles are functionally engaged in the active muscle chain. In the conducted experiments, the activity of chosen muscles was detected by EMG testing and the performance of movements or the quality of posture was measured by the kinematic-dynamic analysis.

3.1.1. Gait

Nowadays, the topic of normal gait is discussed worldwide by academics (mid gait - Young, 1997). It is an activity which is hardly avoided by pregnant women even in the latter stages of pregnancy. In addition, a unified methodology for evaluating gait has not been invented yet and therefore the published data about gait in pregnancy has varied dramatically.

Atkinson (1999) compared 3D analysis of gait among one pregnant and one non-pregnant woman. The gait was recorded on video. The subjects were labelled with markers on the acromion, the most distal rib, trochanter major, epicondylus lateralis femoris, malleolus laterilis, and the navel. The data were evaluated by using Motion Capture software and Motion analysis. The results showed that there were significant differences neither in the lumbar spine curvature (the maximum difference about 10 °), gait speed nor flexion and extension in the hip joint.

Bird et al (1999) observed gait among 25 pregnant women at the beginning of gravidity. The results showed dilatation of the weight-bearing base in pregnancy.

Butler et al (2006) studied the ability of keeping balance and stability. Moreover, it was tested if falling in pregnancy was related to the decreased postural stability. The reason for that was the fact that almost one quarter of pregnant women suffered a fall. The number is comparable with people who are over 65 years old. Twelve pregnant and non-pregnant women (average age 31) took part in the experiment. At the 11th – 14th, 19th – 22nd, and 36th – 39th week of pregnancy and 6 – 8 weeks after birth the markers were placed on the participants and their gait was recorded by a 3D device. The observed parameters stayed relatively the same within both groups of participants. However, both the extension of the hip joint and the flexion of the knee joint increased at the end of the standing phase (this phenomenon is usually guided by greater extension of the knee joint between the half and the end of standing phase). The results also showed no difference in the width of the base and the position of the thorax during walking cycles. The speed of gait was increasing together with the length of steps from the first to the third trimester ($p < 0.05$). There was found no difference in the postural stability between the groups of the women in the first trimester of pregnancy. Furthermore, the women were also tested standing with closed eyes.

In that case postural stability increased in the group of the pregnant women who were in the second or third trimester and even stayed lower after 6 – 8 weeks after birth. In addition, the difference between the groups was directly proportional to the stage of pregnancy.

Another paper was published by Foti et al (2000). The paper described gait of 15 women in the second half of the third trimester and one year after birth. The chosen gait parameters were obtained by the system for 3D motion analysis and the dynamometric platform. The obtained data were compared by using a paired test. The watched parameters were ranges of the joint motions, moments of inertia, and the width of the weight-bearing base. No difference was measured in the speed of gait, length of steps or gait rhythm. Neither the width of steps nor mobility of the pelvis and the ankle joint was significantly changed ($p > 0.05$). Despite the above mentioned facts, an anterior pelvic angel increased about 4° in pregnancy; however, there were considerable variations between the participants (from – 13° to + 10°). In addition to the results, the flexion and adduction of the hip joint largely increased. Finally it was discovered that the phase of double foot-holding increased and the phase of foot swing was shortened.

Golomer et al. (1991) investigated gait with and without a burden. The group of ten pregnant and 20 non-pregnant women carried the burden. The speed of gait and the characteristics of the foot-ground interaction were monitored. The results presented that the speed of gait of the pregnant women did not depend on carrying the burden. The rhythm of gait was faster for the pregnant woman and the length of steps was shorter during pregnancy. The length of steps stayed about the same with or without the burden.

Lymbery a Gilleard (2005) employed an 8-camera system for 3D motion analysis and also measured the pressure of feet to the ground at the end of pregnancy and after birth. They measured 13 pregnant women at the 38th week of gravidity and 8 weeks after birth. They listed a greater width of the weight-bearing base at the end of pregnancy. The mediolateral reaction force on the ground was increasing in the medial direction. The center of pressure (COP) was moved to the centre and anteriorly.

The paper by Osmana et al (2002) discussed 4 pregnant women at the different stage of gravidity and 4 women after birth. Their walking stereotypes were analysed by using the 3D system Peak Motus 2000 and a video camera that took pictures of reflective markers glued to the body. The activity of paravertebral muscles was measured with EMG in the area of lumbar spine (L4/5). Next, the COP was measured on the dynamometric measuring platform Kistler and the interaction forces between feet and ground were analysed in three directions (vertical, lateral, and anteroposterioric). The data of the groups were compared and results were interpreted. The width of the weight-bearing base was increasing in pregnancy. The mean width of the weight-bearing base increased from 168 mm in the first trimester to 350 mm in the third trimester (increase about 50 %). The mediolateral component of reaction force on the platform increased up to 15 % of the body weight.

The experiment conducted in our laboratory was carried out on six pregnant women who were observed during the full duration of pregnancy. Their gait stereotypes were always analysed at the end of each trimester. The kinematic properties were received thanks to the system Qualisys. The system uses infra-sensitive markers and enables one to observe defined spots in time. The dynamometric measuring platform Kistler read simultaneously reaction force between feet and the platform. The placement of the markers is displayed in figure 1.

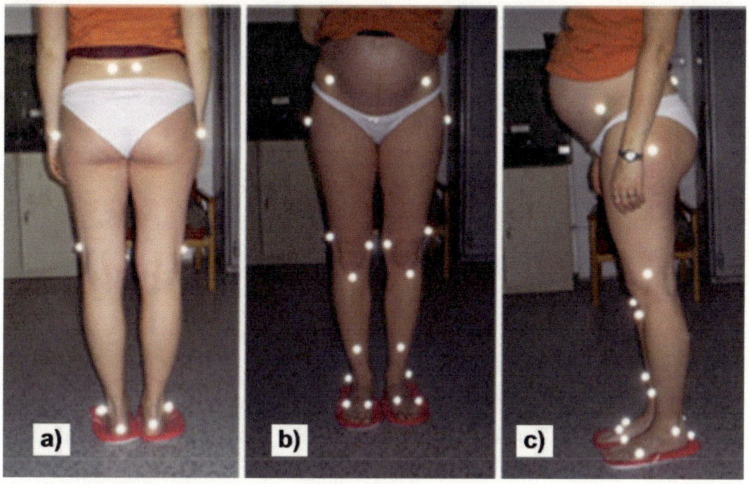

Figure 1. The markers location (a) rear; b) front; c) side.

The observed values were the speed of gait, the weight-bearing base, the time of swing and standing phases, the time of double foot-holding phases, and impulses of the vertical, accelerating, and decelerating forces.

The results are well presented in figure 2. The down-pointed arrow means a decrease in the parameter, the up-pointed arrow means an increase in the parameter and the horizontal arrow symbolizes a steady state. The dash represents no measurement was carried due to birth.

Proband		1	2	3	4	5	6
Gait velocity	1.-2. trimestr	↘	↘	↘	↗	↘	↗
	2.-3. trimestr	↘	↗	↗	↗	↗	-
Supporting base width	1.-2. trimestr	↔	↔	↗	↗	↗	↗
	2.-3. trimestr	↔	↔	↘	↘	↗	-
Swing phase	1.-2. trimestr	↘	↘	↘	↘	↗	↗
	2.-3. trimestr	↗	↗	↘	↘	↘	-
Stand phase	1.-2. trimestr	↗	↗	↗	↗	↘	↘
	2.-3. trimestr	↘	↘	↗	↗	↗	-
Double support phase	1.-2. trimestr	↗	↗	↗	↗	↘	↘
	2.-3. trimestr	↘	↘	↗	↗	↗	-
Vertical force impulse	1.-2. trimestr	↗	↗	↗	↗	↗	↗
	2.-3. trimestr	↗	↗	↗	↗	↗	-
Deceleration force impulse	1.-2. trimestr	↗	↗	↗	↗	↗	↗
	2.-3. trimestr	↗	↗	↗	↗	↗	-
Acceleration force impulse	1.-2. trimestr	↗	↗	↗	↗	↗	↗
	2.-3. trimestr	↗	↗	↗	↗	↗	-

Figure 2. The results of the study.

It is obvious that examined parameters have embodied a high interindividual variability. The variability is strongly related to the current fitness of the women and the foetus position. According to the results only an increase in weight-bearing base has been proven.

3.1.2. Standing

The situation about standing strongly reminds the state of the gait research. The information varies mainly in the area of body posture and the lower back positioning.

Kovalčíková (1990) dealt with curvatures of the spine in sagittal plane and the angle of pelvic among women in the single trimesters of pregnancy, after birth (post partum) and after puerperium (post puerperium). The number of 384 pregnant women was divided into three groups; athletes, women psychosomatically ready to deliver a child, and non-athletes. The depth of neck and lumbar lordosis as well as the angle of the pelvis were examined in standing. The increase of neck and lumbar lordosis was confirmed among all three groups in pregnancy and the state started returning to the normal after birth and after puerperium. The mean angle of the pelvis was decreasing in pregnancy (flexion occurring) and after birth, puerperium the angle was increasing (extension occurring). The most significant changes were listed in the group of non-athletes.

The same results, increasing of lumbar lordosis, were confirmed by Otman et al (1989). In the study, 40 pregnant women were tested. It was written that lumbar lordosis increased significantly in pregnancy. On the other hand it got smaller after birth and it became even smaller at the 6th week after birth but it was still bigger than in the first trimester of pregnancy.

Moore et al (1990) published that the lumbar spine was being flatted and the thoracic spine did not change its shape in pregnancy. For the experiment a special suit was constructed. The suit was covered with ten markers along the thoracic spine between Th1 and L5 and then 25 women were measured form the 16th week of pregnancy to birth and again two months after birth. The side photography was taken of the area of the thorax and the profile of the outer skin was established. The results of that study was that lordosis decreased among 56 % of women at the 16th to 32nd week of pregnancy and after that period lordosis increased among 44 % but it still stayed smaller than the curvatures before pregnancy. Both the kyphotic angle and the position of centre of gravity did not move significantly.

Kušová (2004) conducted a study on 15 women that were examined through the use of Moiré tomography in the second and the ninth month of pregnancy and again at the 7th week after birth. The curvatures in sagittal plane and asymmetries of the trunk were evaluated. The results showed that thoracic kyphosis decreased among four out of six women between the first and third trimester. Lumbar lordosis increased in four women and no change was observed for one participant. There was no change in thoracic kyphosis in two women, in one there was an increase of lumbar lordosis, and in two no change again between the 9th month of gravidity and the 7th week after birth. In the period from the first trimester to the 7th week after birth, thoracic kyphosis increased in two women, decreased in one and did not change in two. Lumbar lordosis increased in two women, decreased in one, and did not change in two participants. The other changes considered as errors were mainly influenced by the variability and instability of standing. The author stated that there was no significant relationship between the changes of the spine shape and pregnancy.

The aim of our study at the field of standing has been focused on finding the objective methodology that scores the changes of mass distribution in the body of pregnant women in comparison with nonpregnant. For this reason side photography segmentation of participants was projected (figure 3).

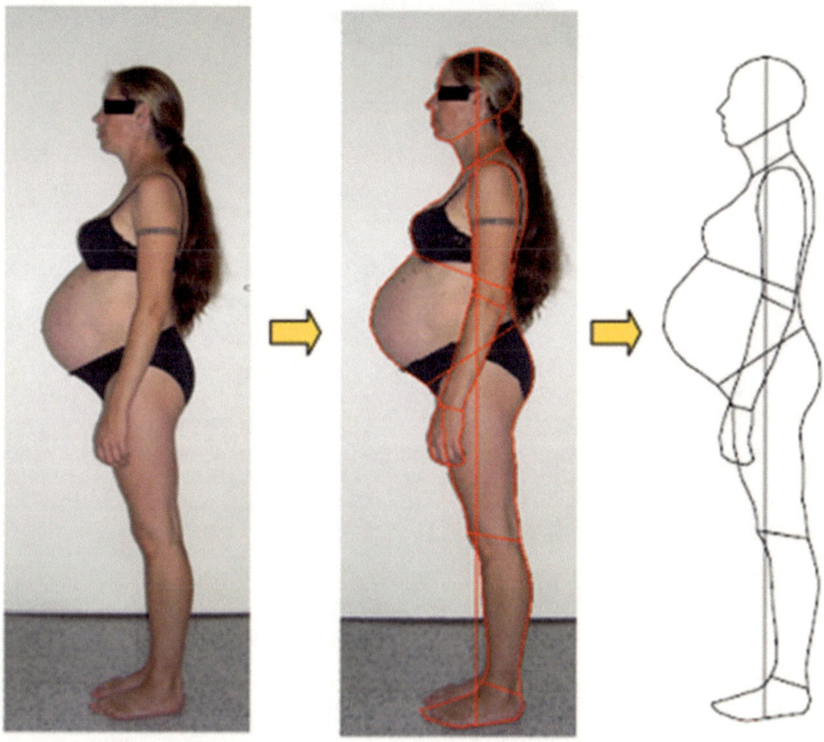

Figure 3. The segmentation process.

The position of the body axis (the line of the centre of gravity), which divides the segments into the front and rear parts, matches the line of action of force of gravity. The force passes through the centre of gravity and it is perpendicular to the ground. Its projection into the ground was established thanks to the dynamometric measuring platform Kistler. The recorded video was used to support the previous experiment. The film showed the marked position of the centre of gravity projection through the use of the dimensions that labelled the relationship between the system of coordinates of the Kistler platform and the participants' ankles (figure 4). The reviewed value in our work was the dimension *c*.

For better orientation, the segmental marking was established topically (figure 5).

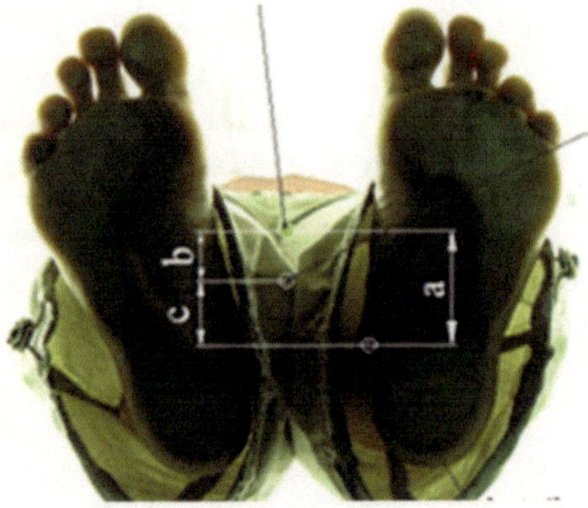

Figure 4. Location of the center of mass projection

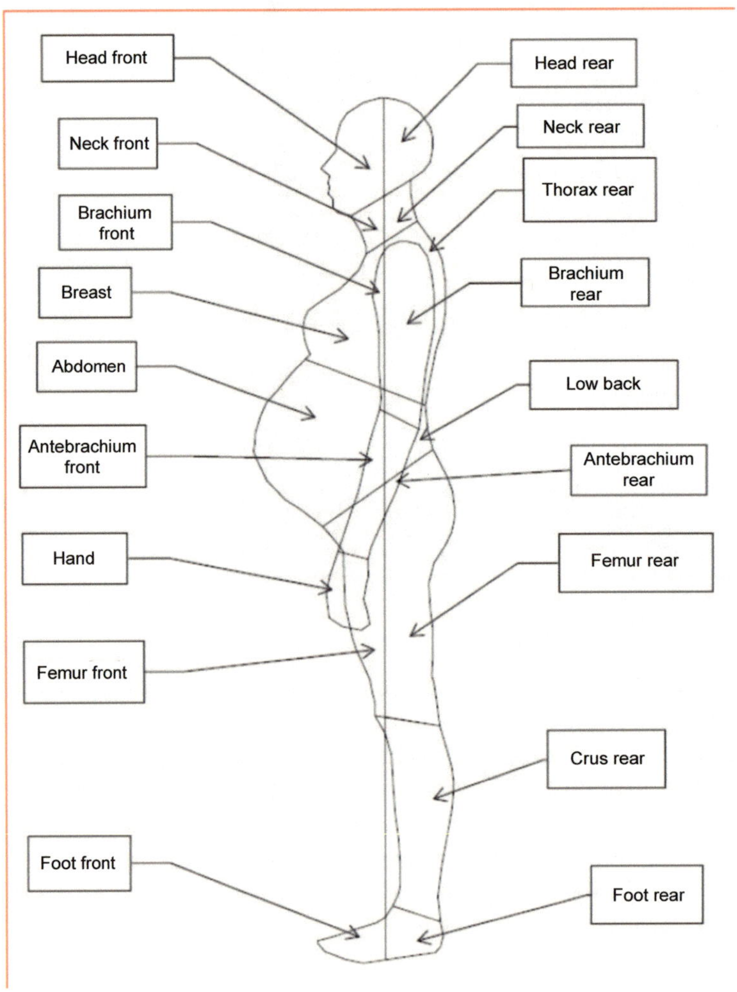

Figure 5. The adopted terminology.

The surface volume of each segment was recounted with the correlation to the weight to avoid data bias.

In the experiment 10 pregnant and 10 non-pregnant women of the same age were tested. The results showed that no significant change in the position of centre of gravity occurred in any direction in pregnancy. The explanation was offered by the analysis of the weights of the body segments. The analysis showed that the progressive state of pregnancy affects growing of the breast, thoracic spine and rear thigh segments up to 4 %. Those changes compensated each other and thus the position of centre of gravity did not differ. According to the analysis of momentum equilibrium it was proved that the momentum impact had forward tendency among the pregnant women. The fact is that the collected data were at the edge of accuracy of the used evaluating methods and therefore it cannot be listed that pregnant women had worse posture stability. The study discovered that pregnancy hardly affects lumbar lordosis and the effect is even smaller among women with a high fitness level.

3.1.3. The dynamic parameters of the gravid abdomen and low back pain

According to the increasing weight of the abdomen in the progressive stages of pregnancy, the inertial effects cannot be neglected or underestimated even during trivial locomotion. The gravid abdomen behaves as an inverted pendulum, which is primarily stabilized by the fibrous suspensory apparatus of the uterus and by the muscles of the abdominal wall. The loading in this area is transferred through the sacrouterinne ligaments to the areas of the low back and lumbosacral junction. This continuous loading consequently leads to overloading of the involved tissue structures which is expressed by pain in the areas mentioned above.

The aim of our research in this field is to establish the influence of changes in dynamic properties of the gravid abdomen and the related force effect on the lumbar region. For changing the aforementioned phenomena, the under mentioned commercially available pregnancy belts were applied in the experiments (figure 6).

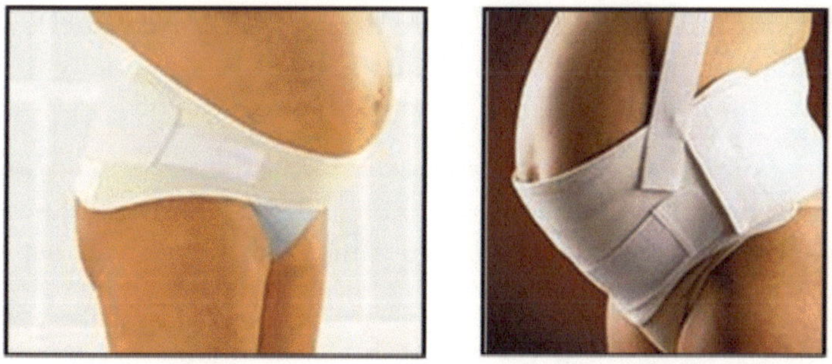

(a) (b)

Figure 6. a) pregnancy belt without braces - Cellacare Materna (www.lohmann-rauscher.cz) b) pregnancy belt with braces - Materna (www.ergon.cz)

For the acquisition of the kinematic data, the Qualisys system was used. The force (dynamic) effects were detected by the Kistler equipment. The experiment was conducted on two pregnant women in the third trimester.

In the first phase of the experiment, the normal gait was analyzed. The analysis was focused on the movements of the marker that was placed on the navel in cranio-caudal and latero-lateral direction (figure 7).

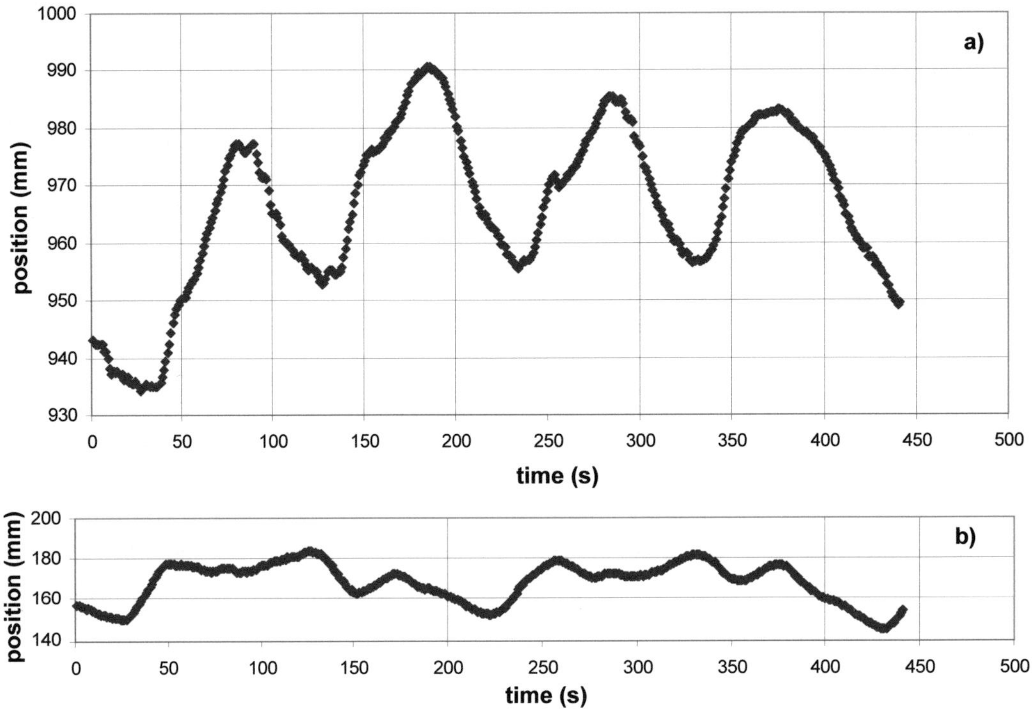

Figure 7. The navel motion (a) caudo-cranial; b) latero-lateral.

The data evaluation was based on mutual comparison of the displayed curves for the measurements without the belt, with the belt and with the belt and braces for both participants. The observed phenomenons were the significant frequencies characterized by the highest amplitudes. The results showed that the belts had a totally negligible effect in this respect, because the change of those frequencies was not found.

In the second stage of the experiment, the vibrations of the participant's gravid abdomen were observed after the fall on heels after standing on tiptoe. The caudo-cranial movement of the navel marker was recorded. The evaluation was performed separately for each direction (figure 8).

The last stage of the experiment contained a questionnaire investigation which was designed to explore the participant's feelings about the belts and the connection between the lumbar pain and wearing the belts. In the final part 11 pregnant women in the third trimester participated. The selected belt type was worn for 14 days except for sleeping.

The obtained results confirmed the reduction of pain in the observed area from 20 up to 76%.

According to the results, the importance of the supporting devices is mainly to decreased loading in the stressed areas and reduce the utilization of the involved tissue structures.

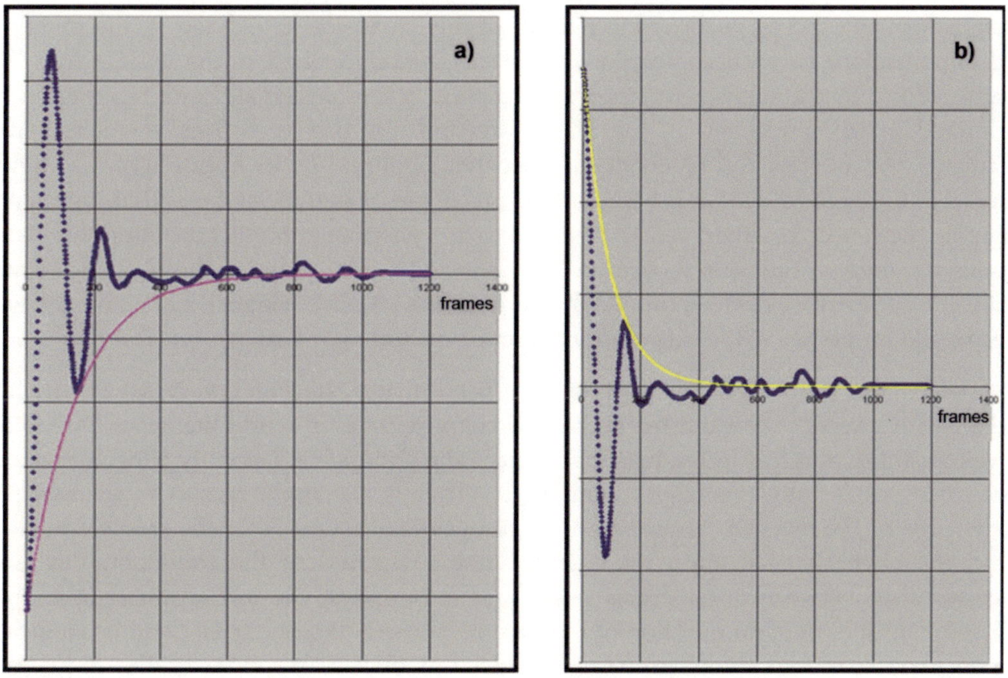

Figure 8. The navel motion suppressing a) downward direction; b) upward direction.

3.2. Endopelvic fascia

The endopelvic fascia is the soft tissue surrounding the vagina. It is attached to the pelvic walls and supports the pelvic viscera - urethra, bladder, cervix, uterus and rectum. Because the fascia is a relatively shape-complicated organ and its various parts are exposed to different mechanical loading, it can be reasonably assumed, that their mechanical properties will vary according to the appropriate field. Regarding the complex structure of the endopelvic fascia, some strength tests through its whole length are difficult to perform. The research is then focused on the areas where the fascia is relatively accessible and where some of its parts can be removed during standard surgeries without causing any inconvenience for patients. The main monitored parameters are elasticity and viscosity, which are represented by the identifiable proteins (e. g. collagen, elastin, etc.) and their mutual arrangement.

Our current work has mainly targeted the issue of long-term postnatal complications in terms of biomechanics, which are largely caused by the processes occurring during birth. The specific goal of the research was the endopelvic fascia and its properties in relation to its intimate relationship to the vaginal mucosa. The changes occurring during birth are also characterized by the greater or minor damaged of tissues. It may also results in a functional failure of the pelvic floor. The damage usually has a multifunctional character and also diverse consequences, however, they are never beneficial for the health of the patient.

The birth is initiated by uterine activity which leads to the gradual extending of the lower uterine segment and cervix. The mechanism of the expansion is allowed by the muscular cell organization. At each contraction the uterus is straightened to the middle line. The uterus is fixed by the suspensory apparatus (especially uteroingvinal chorda) so the fundus is limited in its movement. In the distal direction, the uterus is fixed by sacrouterine ligaments, the muscles and ligaments of the pelvic floor and by its insertion of the vagina. Thanks to the experience that is based on the above mentioned facts, the birth duration and complications, and the other well-known factors it is possible to predict the injury of related tissues and organs. The main recognized causes include injuries such as problematic vaginal birth, chronic increase of the intra-abdominal pressure (obesity, coughs), aging and changed mechanical properties of the suspensory apparatus including the endopelvic fascia.

The mechanical properties of the fascia have been investigated only very marginally and there is still a lack of the valid biomechanical characteristics in world literature. Due to the development of surgical techniques that replace the endopelvic fascia by allogen implants that often result into over rigid spare septa. That is the main reason to increase the knowledge of the mechanical properties of autogenous tissues. From the medical point of view, the biomechanical approach is irreplaceable. Because of the continuing "material disagreement" between the operated tissue and the implant, the foreign material is often refused, which is rather a question of immune response and this can be pharmacologically suppressed. A more serious problem is often the unclear response of the implant to mechanical loading. This is the main factor that influences the success of the surgery, because complicated thermo-visco-plasto-elastic properties of living tissues cannot be substituted by a purely mechanical replacement.

Within the latest phase of our research, 16 samples of vaginal wall with fascia were measured by standardized uni-axial tensile test to determine their "referenced" properties. Next, 6 samples of the implants were measured by the same procedure. The following text presents the proposed and used methods of processing and evaluating of the measured data. At the end, the obtained findings associated with the monitored parameters such as pregnancy, number of completed pregnancies and age of the donor women are listed.

For description of observed materials, we used the linear elastic modulus, which is defined by following formula:

$$F = K \, . \, \Delta l, \tag{1}$$

where $\boldsymbol{K}$ is stiffness (rigidity) (N/mm), $\boldsymbol{F}$ force (N) and $\boldsymbol{\Delta l}$ relative extension (mm).

Regarding the real organization of both tissue structures in the samples (figure 9), we created a complete model of the tested samples by parallel junction of two rigidities, which can be described by the following equation of the force balance:

$$F_{FP} = F_P + F_F, \quad (2)$$

where F_{FP} is the force detected by the measuring head of the device, F_P is the reaction force given by properties of the vaginal wall and F_F is the force from endopelvic fascia.

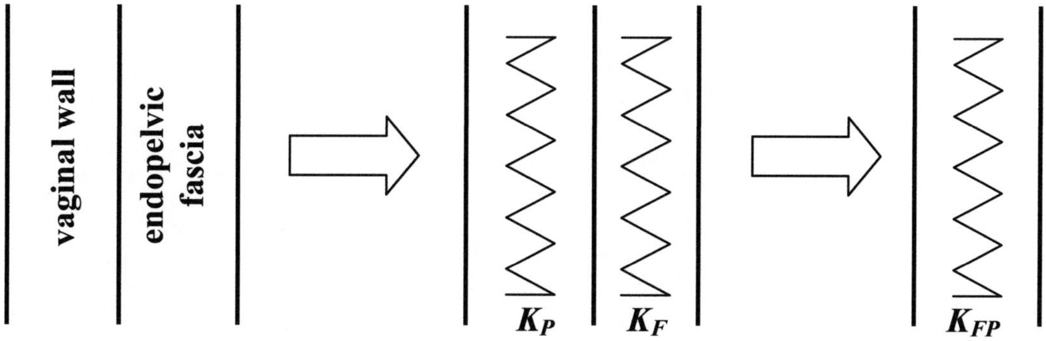

Figure 9. The tissue structure layout chart.

Using the formula (1), the equation (2) can be arranged to the next shape:

$$K_{FP}.\Delta l = K_P.\Delta l + K_F.\Delta l, \quad (3)$$

and after rearrangement, the relation for rigidity of the vaginal wall is obtained:

$$K_P = K_{FP} - K_F \quad (4)$$

Regarding the data obtained from the performed experiments, this relationship can be used to calculate the rigidity of the vaginal wall at the moment of its rupture, when the rigidity of the separated endopelvic fascia is known.

For each dependency between the force and extension (figure 10), several particular magnitudes of the rigidity of the used model were obtained.

The yellow marked area in figure 10a is the record of the cyclic "preload" of the sample in order to stabilize its mechanical properties. The slight vacillations of measured curves (figure 10b, red marked area) showed that the prolongation without the further presumed force increase may be interpreted e.g. as moments, where some minor damages had happened in the tissue without influence on overall stability of tested sample's response. The major breakthrough in the sample response's course was the vaginal wall rupture (figure 10b, yellow marked area). The following graph course (figure 10b, area 8 and 9) was then formed only by the endopelvic fascia response. The moment of the vaginal wall rupture and also endopelvic fascia rupture was well detectable even on the synchronous video recording of the experiment.

The curve (figure 10b) was further divided into the particular sections with a linear character, which were assigned rigidities characterizing the vaginal wall with endopelvic fascia as a whole (figure 10b, areas 1 to 7) and rigidities of the endopelvic fascia separately (figure 10b, areas 8, 9). Applying the above formulas, the rigidity of the vaginal wall can be calculated.

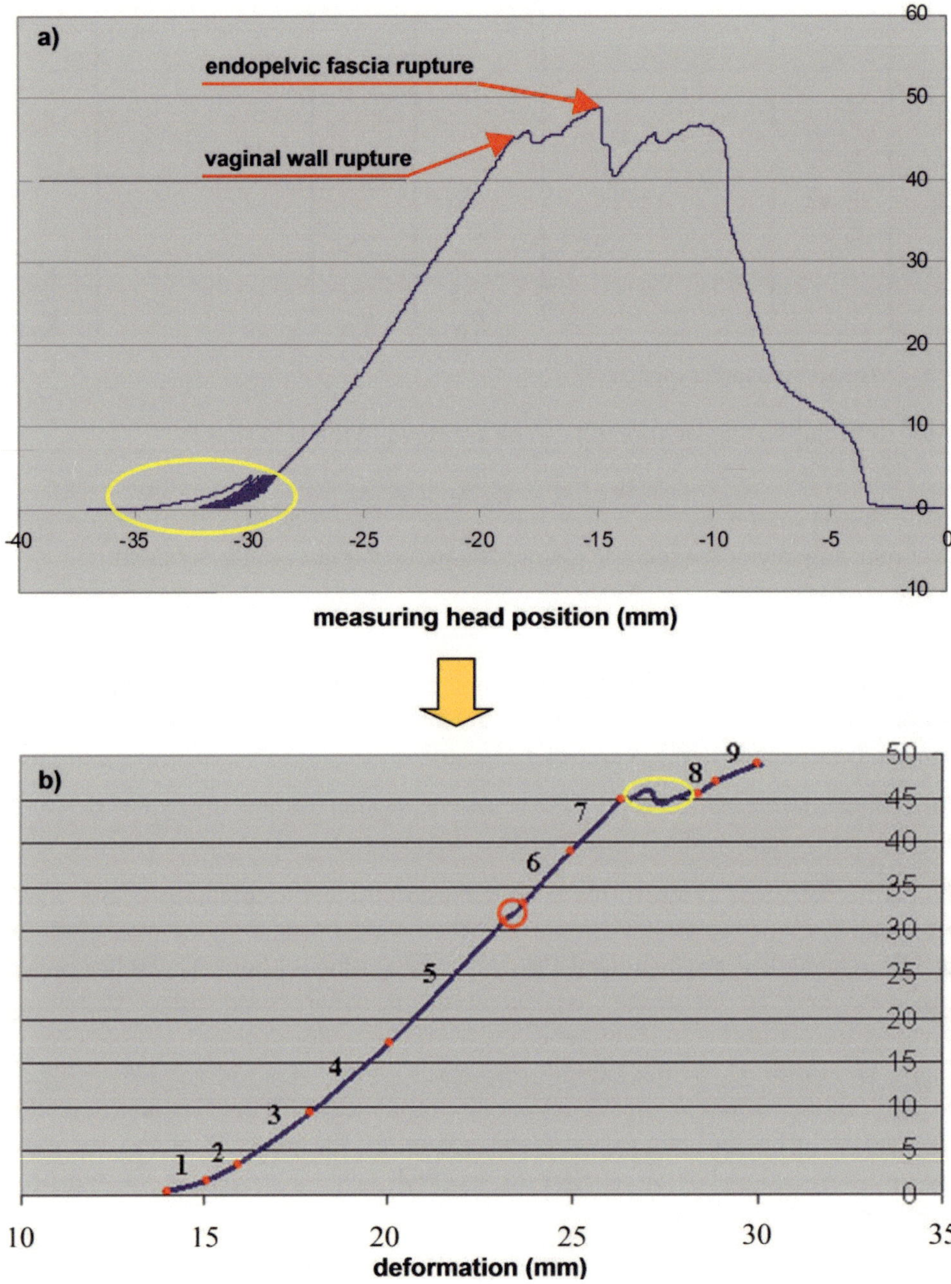

Figure 10. Measurement record (a) and evaluated section (b).

According to the measurement curve analysis and comparison of calculated rigidities the following can be stated:

1. The vaginal wall endures lesser prolongation compared to the endopelvic fascia. This conclusion is valid for all our experiments performed so far, independently on patient anamnesis.
2. Samples rigidity increases with deformation and after reaching maximum decreases while heading for the rupture (figure 11). The curve has a concave characteristic and it is visible on all tested samples.

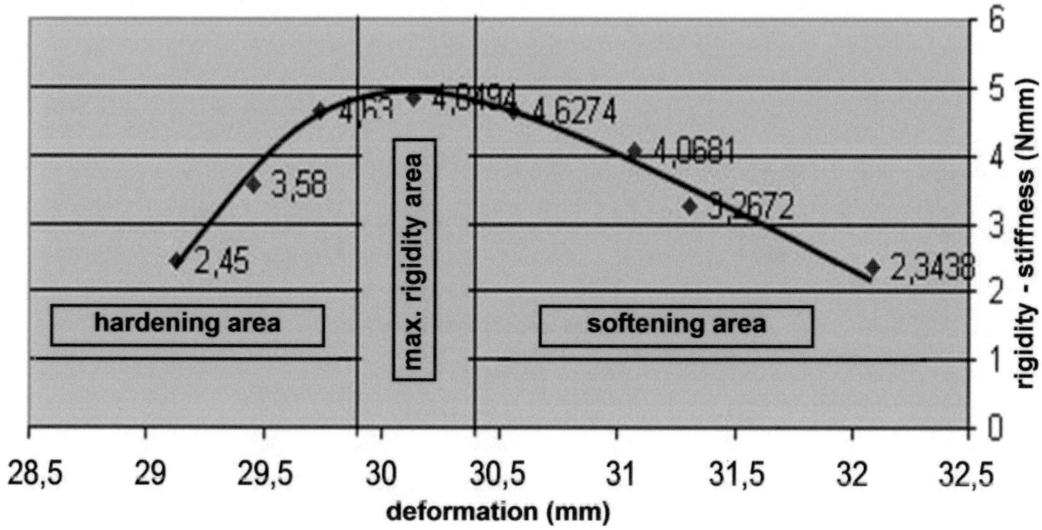

Figure 11. Rigidity – prolongation relation of fascia + vagina complex (an example).

3. After the vaginal wall rupture the rigidity of the endopelvic fascia decreases with increasing deformation. This decrease can be considered linear with satisfying precision.

From the current results it can be concluded that the endopelvic fascia has relatively stable properties that are changed significantly only in pregnancy and stabilized again after it. In terms of long-term changes associated with a decrease of mechanical properties of the fascia the crucial parameter is the age of a woman. The number of completed pregnancies exhibits no significant influence.

The processing and evaluating of the data from the second phase of the experiment corresponded to the methods described above. The data were arranged into graphs (figure 12) and the dependence of rigidity on extension of the samples was evaluated.

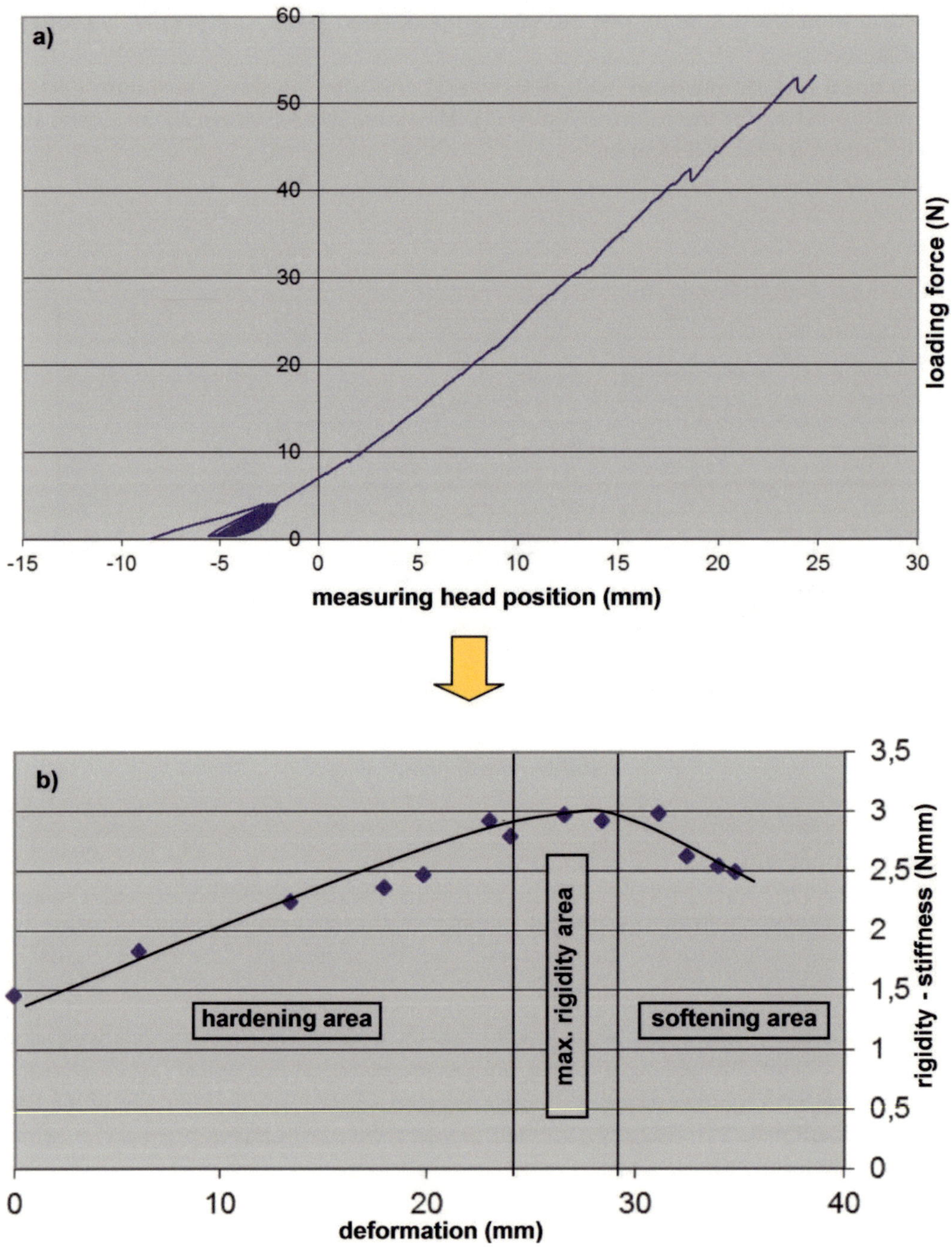

Figure 12. a) Measurement record and rigidity; b) prolongation relation of the implant.

The comparison of the graphs 12a and 12b shows that the response of samples of vaginal wall with the endopelvic fascia and samples of used implants is similar. The question is, whether these values of the implants rigidity are convenient for their purpose. A reliable answer to this question tasks for an extensive study, however, it must be fulfilled that the implant should compensate for the differences between rigidity of the healthy and damaged tissues.

Author details

Karel Jelen, František Lopot, Daniel Hadraba and Martina Lopotova
Charles University, FSPE, Department of Anatomy and Biomechanics, Prague, Czech Republic

Hynek Herman
Institute of the Care of Mother and Child, Prague, Czech Republic

Acknowledgement

This work has been supported by grants from the Grant Agency of the Czech Republic provided for the period 2010-2013.

4. References

Atkinson, B., Stirling, C., Sukhtankar, A. *Gait Differences Between Pregnant and Non-pregnant Women* [on-line]. ©1996, last revised 9/99 [cit. 2008-03-20].

Bird, A.R., Menz, H.B., Hyde, C.C. The effect of pregnancy on footprint parametres. A prospective investigation. In *Journal of the American Podiatric Medical Association,* 1999, vol. 89, no. 8, p. 405-409.

Butler, E. et al. An investigation of gait and postural balance during pregnancy. In *Gait & Posture,* 2006, vol. 24, no. 2, p. S128-S129.

Foti, T., Davids, J.R., Bagley, A. A Biomechanical Analysis of Gait During Pregnancy. In *Journal of Bone and Joint Surgery,* 2000, vol. 82-A, no. 5, p. 625-632.

Golomer, E., Ducher, D., Arfi, Gs., Sud, R. A study of pregnant women while walking and while carrying a weight. In *Journal de Gynecologie Obstetrique et Biologie de la Reproduction,* 1991, vol. 20, no. 3, p. 406-412.

Kovalčíková, J. *Dynamika chrbtice a statika panvy žien počas fyziologickej gravidity.* Bratislava : Univerzita Komenského v Bratislavě, 1990. ISBN 80223-0208-2.

Kušová, Sabina. *Dynamika vybraných parametrů axiálního systému gravidních žen a žen do jednoho roku po porodu.* Praha, 2004, 230 s. Disertační práce na FTVS UK, Katedra anatomie a biomechaniky. Vedoucí práce Doc.Karel Jelen, CSc.

Lymbery J.K., Gilleard, W. The Stance Phase of Walking During Late Pregnancy. In *Journal of the American Podiatric Medical Association,* 2005, vol. 95, no. 3, p. 247-253.

Moore, K., Dumas, G.A., Raid, J.G. Postural changes associated with pregnancy and their relationship with low-back pain. In *Clinical Biomechanics*, 1990, vol. 5, no. 3, p. 169-174.

Osman, N.A., Ghazali, M.R. Biomechanical evaluation on gait patterns of pregnant subjects. In *Journal of Mechanics*

Chapter 2

Locomotion Transition Scheme of Multi-Locomotion Robot

Tadayoshi Aoyama, Taisuke Kobayashi, Zhiguo Lu, Kosuke Sekiyama,
Yasuhisa Hasegawa and Toshio Fukuda

Additional information is available at the end of the chapter

http://dx.doi.org/10.5772/49997

1. Introduction

There are researches aiming to give a high environmental adaptability to robots. Until now stable locomotion of robots in complex environment such as outside rough terrain or steep slope has been realized [1–7]. Locomotion in the most of researches adapted to complex environment has been realized by single type of locomotion form. On the other hand, we have proposed Multi-Locomotion Robot (MLR) that can perform several kinds of locomotion and has high mobility as shown in Fig. 1 [8]. By using MLR, we have realized independently biped and quadruped walking, brachiation, and climbing motion so far [9–15]. Next research issue of MLR is to develop a systematic transition system from one locomotion form to the other.

Aoi et al. proposed transition motion from biped to quadruped walking by changing the parameters of the nonlinear oscillator and conducted experimental verification [16, 17]. These works focuse on realization of a stable motion transfer and the transition according to external environment has not been discussed. Meanwhile, Asa et al. discussed the dynamic motion transition using the bifurcation of control parameters and realized motion transition between biped and quadruped walking [18]. These conventional researches aimed to realize a motion transfer between biped and quadruped walking. The transition motion of control system is constructed by using the Central Pattern Generator (CPG); the motion transfer of is realized by attractor transfer mechanism.

On the other hand, we aim to select suitable motion pattern for robots based on external environment and internal state of the robots and realize motion transfer from current motion to the other. In this chapter, we focus on biped and quadruped walking as motion patterns and report the suitable motion selection between biped and quadruped walk considering the walking stability and efficiency. Motion and recognition uncertainty is focused as factors to effect a realization of walking; then walking stability is evaluated from stability

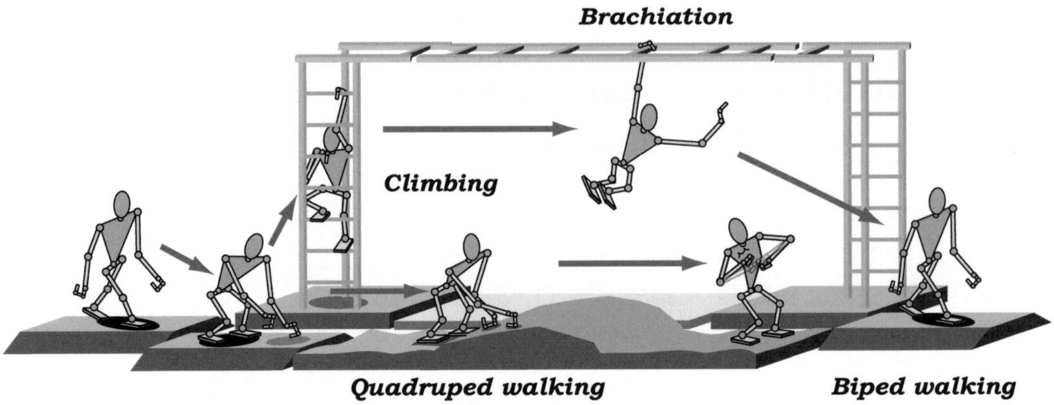

Figure 1. Concept of Multi-Locomotion Robot

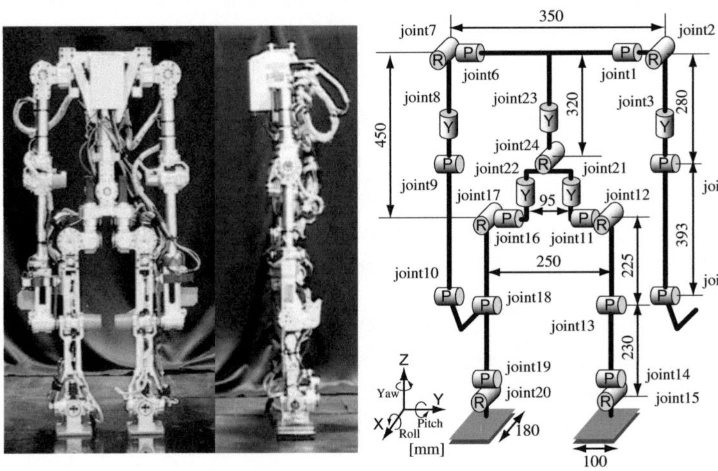

Figure 2. Gorilla robot Ⅲ

evaluation parameters that have multiple uncertainties. Since dimension or class of the stability evaluation parameters that have uncertainty are different and the parameters cannot be used with uniformity, the parameters are integrated into the risk of falling down as the belief with Bayesian Network. The internal model to select the optimized motion pattern that minimizes falling down risk and maximizes the transfer efficiency is designed. Finally suitable locomotion selection between biped walking and quadruped walking is experimentally realized.

2. Multi-locomotion robot

2.1. Gorilla robot III

Multi-Locomotion Robot is a novel bio-inspired robot which can perform in stand-alone several kinds of locomotion such as biped walking, quadruped walking, and brachiation. We

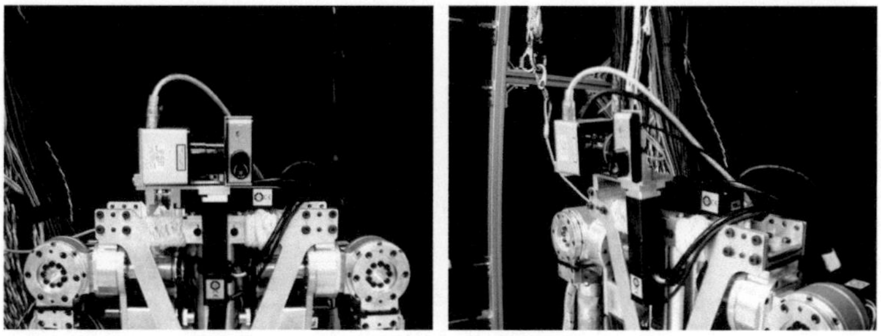

Figure 3. Laser range finder

built and developed Gorilla Robot III as a prototype of Multi-Locomotion Robot [8]. Overview and link structure of Gorilla Robot III is shown in Fig. 2. Its height is about 1.0 [m] and weight is about 24.0 [kg]. The mechanical structure is designed as follows: 6 DOF leg, 5 DOF arm, 2 DOF lumbar. Each joint is actuated by AC servo motor. Computer, AD/DA board, counter board, and power are set outside the robot.

As a sensor for recognition of slope, a laser range finder is installed at the neck of the robot (see Fig. 3). Its angular resolution is 0.36 [deg], scan angular range is 240 [deg], scan time is 100 [ms], and maximum range of detection is 4.0 [m]. The rotation axes of motors are pitch and yaw axes. In addition a web camera is also installed next to the laser range finder.

2.2. Locomotion mode

In this chapter, we model the robot as a 3D inverted pendulum same as the work for biped walking [19]. The supporting point of the pendulum is assumed to be point-contact. Then, only the heeling force f and the gravity act on Center of Gravity (COG). In this chapter, we use crawl gait as a quadruped walking [14]. In this gait, the idling leg changes, left rear leg, left front leg, right rear leg, and right front leg, in that order. It is designed in order that three feet always contact the ground, COG moves within the triangle which is formed by the three supporting feet. The transition from biped to quadruped posture is made keeping static balance. Before transiting the posture between biped and quadruped stance, the robot stops walking.

3. Locomotion stabilization

3.1. Internal model

In order to realize a robust robotic locomotion in any environment, two abilities are required: planning of the suitable motion based on the recognition of moving environment, and evaluation of generated motion. Then we propose the internal model based on a prediction and feedback as shown in Fig. 4.

Prediction for locomotion plans the locomotion form based on environmental information. Environmental information is sensed by a laser range finder; then the robot determines the suitable gait for the environment. In this research, biped and quadruped walking is focused as the gaits. The robot selects biped walking in the environment that is easy to walk such as flat terrain. Meanwhile the robot selects quadruped walking in the environment that is difficult to walk in biped state such as slope or rough terrain. Also, the robot plans the walking steps and landing position of the selected gait based on recognized terrain. Previously, we designed this prediction for locomotion [20].

The feedback for locomotion evaluates walking stability based on internal condition of the robot. In this chapter, we propose the method of estimating the risk of falling down using Bayesian Networks (BN). In estimating the risk, we set "Robot Model Reliability (Reliability of Internal states)" and "Environmental Model Reliability (Reliability of External dynamics)". Reliability of a robot model shows how far difference between reality motion and locomotion algorithm is, or physical abilities of robot. For example, if the robot has motor trouble, this is low and the risk of falling down is high. Reliability of an environmental model shows how accurately a robot recognizes environment. If robots move in dark, it does not get information of environment, so this parameter is low and the risk of falling down is high. In biped and quadruped walking, the robot evaluates both reliabilities, estimate the risk of falling down and attain an optimum gait adapting to the environments or the conditions. This feedback for locomotion is explained in the next section.

3.2. Stabilization based on internal conditions

3.2.1. Estimation of falling down risk

In this chapter, we consider the uncertainty caused by motion and recognition as the factor of realization of locomotion. Approximation of motion algorithm is pointed out as uncertainty caused by motion. Most robots have models to simplify calculating dynamics. Thus, this gives robot systems uncertainty because there are difference between a reality robot shape and a robot model. Uncertainty caused by recognition is accuracy of sensors, effective ranges of sensor or abstraction of environment. There are many kinds of uncertain parameters which have various dimensions, so it is difficult to deal with them uniformly. Then, these parameters are integrated into the risk of falling down as belief with Bayesian Network. The Bayes theory assumes that parameters have distributions individually, and posterior probability is induced formally by conditional probability. Bayesian Network is the model which describes relations

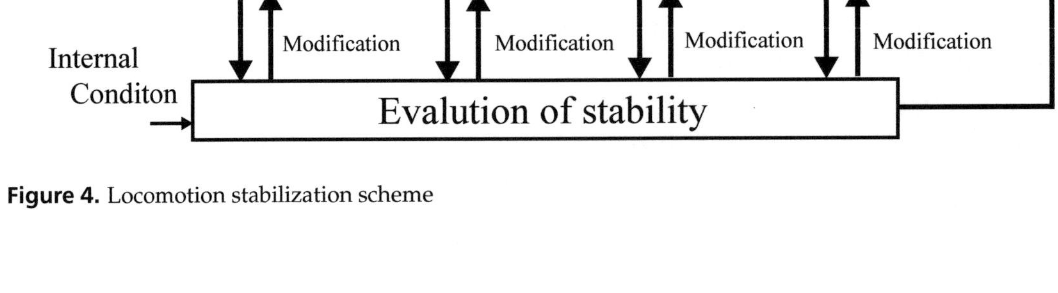

Figure 4. Locomotion stabilization scheme

among phenomenon using probability. We describe the causality between the risk of falling down and the uncertain parameters.

In this research, Bayesian Network shown in Fig. 5 is used to estimate the risk of falling down. First, Bayesian Network estimates Robot Model Reliability "*R*" and Environmental Model Reliability "*E*". Reliability of a Robot Model *R* show how ideal the robot motion is,

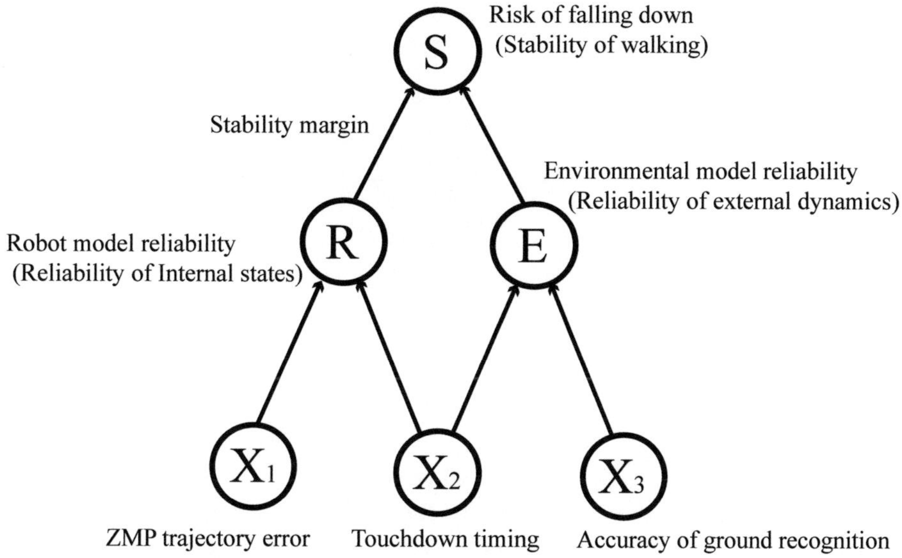

Figure 5. Bayesian Network for locomotion stabilization

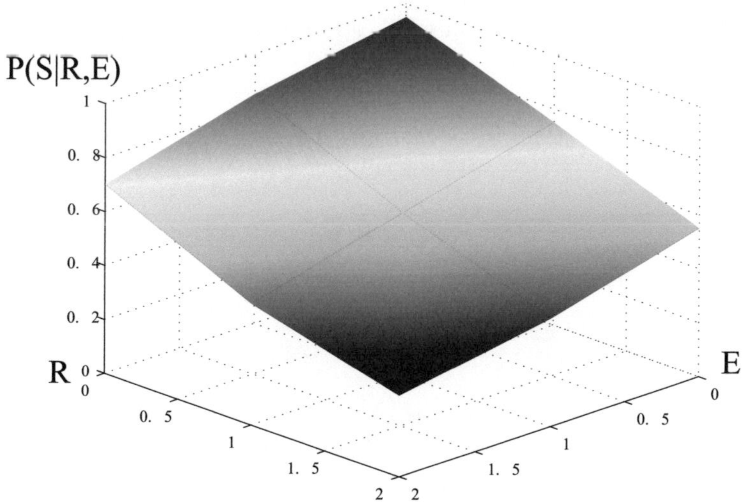

Figure 6. Probability for Biped Walking

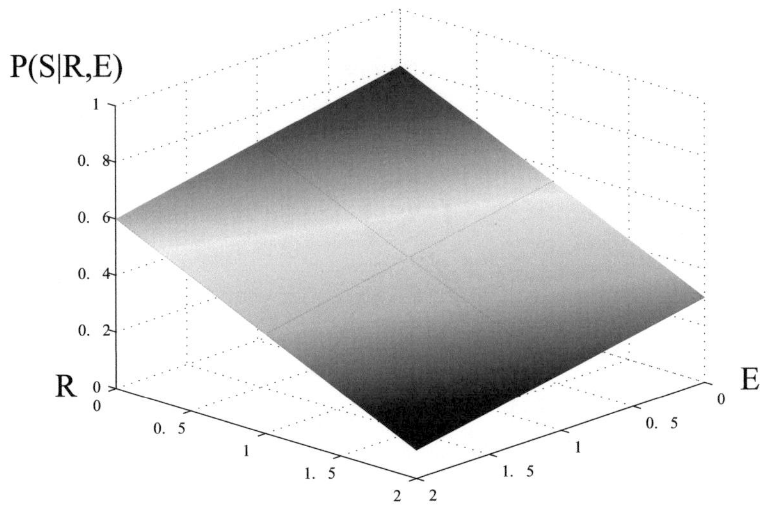

Figure 7. Probability for Quadruped Walking

and describes the capacity of moving. Reliability of an Environmental Model E is an index which shows how correctly the robot perceive the dynamics between the environment and the robot. Secondly, R and E are induced the risk of falling down "S". "$S = 1$" shows falling down, and "$S = 0$" shows not falling down. Probability variables R and E have classes 0, 1, 2 in more reliable order. Then conditional probability $P(S \mid R, E)$ reflects the performance of the robot, and the designer arranges this probability subjectively. Probability distribution of biped walking is different from quadruped walking so that $P(S \mid R, E)$ of biped walking is higher than quadrupled one. Fig.6 and Fig.7show $P(S \mid R, E)$ of biped walking and quadruped walking respectively. The evaluating parameters X_1, X_2, X_3 shown below are observed at real time. Then probability variables from 0 to 4 based on uncertainty which the parameters have input the Bayesian Network. When the probability variable is 0, the situation is most stable. The calculation of Bayesian Network uses the enumeration method shown by (1).

$$P(S=1) = \frac{\sum_{R=0}^{2}\sum_{E=0}^{2} P(S=1, R, E)}{\sum_{S=0}^{1}\sum_{R=0}^{2}\sum_{E=0}^{2} P(S, R, E)}$$

$$= \frac{\sum_{R=0}^{2}\sum_{E=0}^{2} P(S=1 \mid R, E) P(R \mid X_1, X_2) P(E \mid X_2, X_3)}{\sum_{S=0}^{1}\sum_{R=0}^{2}\sum_{E=0}^{2} P(S \mid R, E) P(R \mid X_1, X_2) P(E \mid X_2, X_3)} \tag{1}$$

The evaluating parameters X_1, X_2, X_3 are always observed, so each probability $P(X_1), P(X_2), P(X_3)$ is set 1.

3.2.2. *COG trajectory error* X_1

The position of the center of gravity is measured by the force sensor which the robot put on its four legs. In biped posture, outputs which come from the sixth axis force sensor makes ZMP. In quadruped posture, the center of gravity is calculated with the equilibrium of moments. Then the errors between the desired trajectory and the observed trajectory decides the probability variable X_1.

3.2.3. *Touchdown timing* X_2

The touchdown timing shows differences between the landing and the ground surface actually. When the robot is thrown off balance, or when the recognition is inadequate and the ground is higher than measured point, then the touchdown timing is earlier than the planed timing. In the robot moving, the probability variable X_2 is renewed at every landing.

3.2.4. *Accuracy of ground recognition* X_3

This parameter evaluates the performance of the recognition which the robot has. This shows how much information the robot attain with some sensors, and how abstracted the environmental model which the robot has is. The laser range finder has effective ranges, so over this ranges there is much uncertainty. Then the two-dimension recognition and the approximate algorithm have the uncertainty.

3.3. Consideration of stability margin

The conditional probability $P(S \mid R, E)$ describes the influence which Reliability of a Robot Model R have with the Risk of falling down S. Then when the stability margin is enough large compared with the COG errors, the influence is little even if R goes down. In reverse, when the stability margin is small, R has a big influence on S. Therefore $P(S \mid R, E)$ is decided based on the stability margin. For example, a stability margin in biped posture is smaller than one in quadruped posture, so $P(S \mid R, E)$ in biped posture is larger than in quadruped posture.

3.3.1. *Consideration of stability margin*

The conditional probability $P(S \mid R, E)$ describes the influence which Reliability of a Robot Model R have with the Risk of falling down S. Then when the stability margin is enough large compared with the COG errors, the influence is little even if R goes down. In reverse, when the stability margin is small, R has a big influence on S. Therefore $P(S \mid R, E)$ is decided based on the stability margin. Thus, $P(S \mid R, E)$ is changed by designing the revised value of conditional probability $\Delta P(S \mid R, E)$ shown in Fig. 8 according to the stability margine as follows:

$$P(S \mid R, E) = P(S \mid R, E) + \Delta P(S \mid R, E), \tag{2}$$

$$\Delta P(S \mid R, E) = -\frac{2\Delta P}{k_{max}} k + \Delta P, \tag{3}$$

$$0 \leq k \leq k_{max}, \tag{4}$$

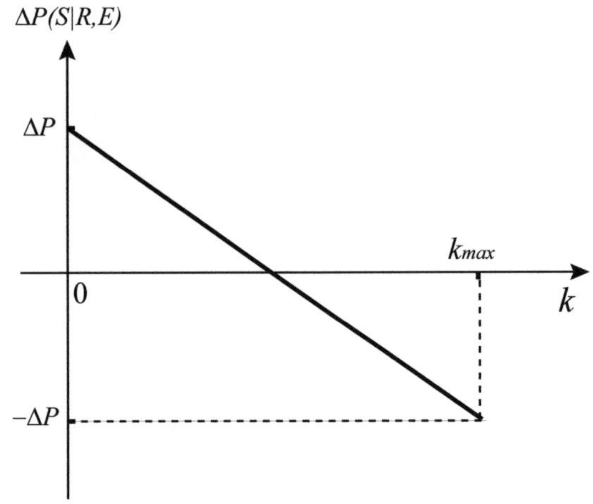

Figure 8. Revised Probability Value According to Stability Margin.

where ΔP is the maximum revised value of conditional probability and k_{max} is the maximum stability margin.

3.3.2. Switching of locomotion mode

The evaluating parameters X_1, X_2, X_3 are observed at real time, and the probability of falling down is estimated. The conditional probabilities used in Bayesian Network are arranged by the subjective judgments of the designer. Therefore, when the robot falls down, the probability of falling down is not always 1.0. So we pay an attention to the fluctuation of the probability. That is, when the robot move in biped posture and the risk of falling down increases, then it has the transition motion from biped to quadruped posture and go quadruped walking. Contrarily the risk decreases in quadruped walking, the robot stands up and go biped walking.

4. Experiments

4.1. Experimental conditions

In this experiment, the robot measures the landform with the laser range finder at starting point, and in walking, it get the gait based on the risk of falling down estimated by Bayesian Network shown in Fig. 9. When the risk is more than β (0.7) in biped posture, the robot squats to get quadruped posture. And when the risk is less than α (0.3) in quadruped posture, it standups. Then the robot in biped posture has three patterns of biped walking a_1, a_2, a_3 which have different efficiency. If the risk decreases, the robot get more efficient gait. In this research, this efficiency is the walking velocity, then a_1, a_2, a_3 are respectively 8.67, 6.67, 4.67[cm/sec] acquired by stride widths changed and the quadruped walking velocity is 3.00[cm/sec]. Both the standup motion and the squat motion take 10[sec] to action. Modifications of its gait are conducted in every walking cycle. The robot aims at minimizing the risk and maximizeing the efficiency all the time.

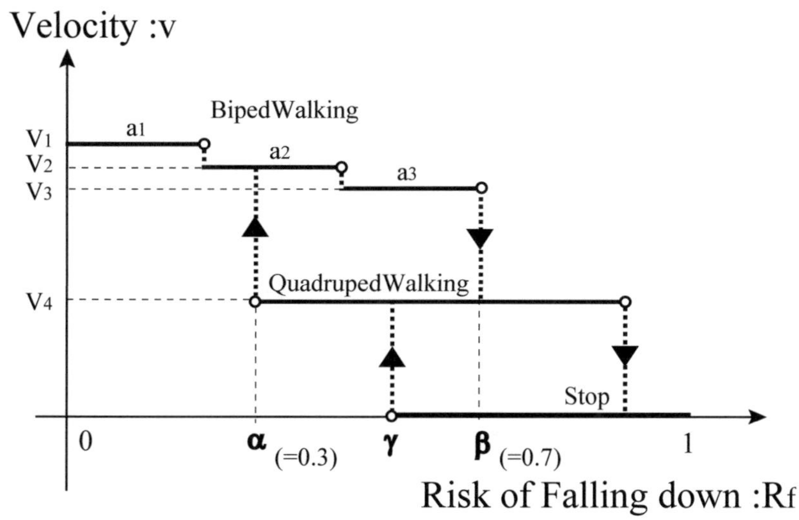

Figure 9. Velocity - Risk of falling down

4.2. Experimental results

4.2.1. Experiment 1: gait selection based on falling down risk (biped to quadruped)

In this experiment, the robot walks on rough ground. There are inequalities which have the maximum height, 5[mm]. This is not recognized by the robot on purpose. We confirmed whether the robot in biped posture changes the gait to quadruped mode because the risk increases.

Fig. 10 shows results about the COG trajectories come from the force sensors. And the COG trajectories induce X_1 shown in Fig. 11. Fig. 12 describes the probability variable X_2. The numbers in these figures are the threshold to apportion the probability variable. In this experiment the node X_1, X_2 have $0, 1, 2, 3, 4$ as the probability variables. When the probability variable is 4, the robot almost falls down. The node X_3 is always 0 because the robot move within the effective ranges of the laser range finder in this experiment. Thus, Fig. 13 is the risk estimated by Bayesian Network. In the transition motion, the risk is 0.0. We can see the transition caused by the risk increasing. Before the robot conducts a squat, the risk is more than β (0.7). And snapshots of the experiment is shown in Fig. 14.

4.2.2. Experiment 2: gait selection based on falling down (quadruped to biped)

The experiment 2 confirms the transition of locomotion form when the robot starts walking in quadruped state and is given shaking disturbances made by human. Fig. 15 shows the estimated risk of falling down derived from the same way in the experiment 1. The risk of falling down is set 0 during transition from quadruped to biped walking. The risk of falling down is temporarily increased due to the shaking disturbances from human. It is confirmed that the robot stop and selects biped walking as locomotion form after disturbances stopped and the risk is less than $\alpha(0.3)$. Fig. 16 shows the snapshots of the experiment 2.

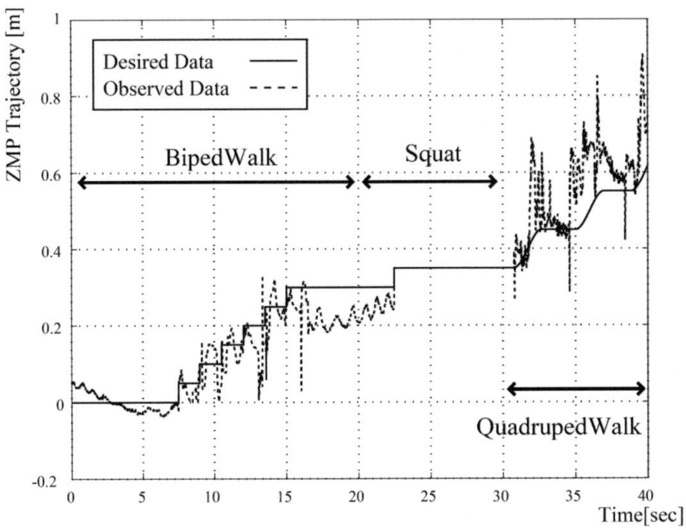

Figure 10. Comparison between desired and actual ZMP trajectory

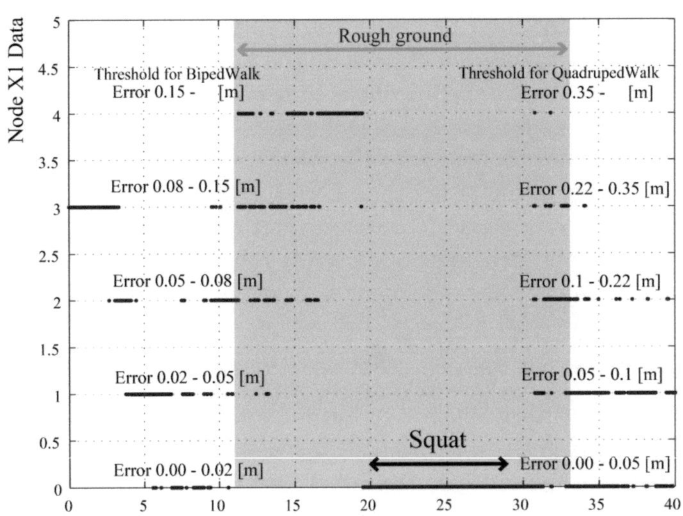

Figure 11. Experimental data of node X_1

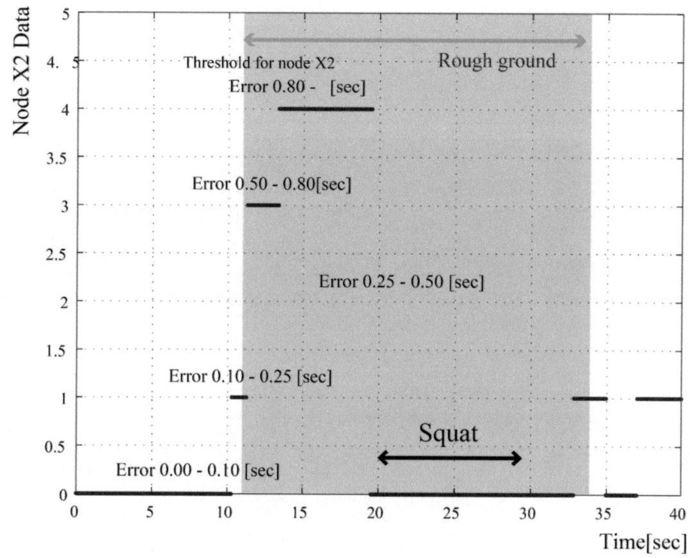

Figure 12. Experimental data of node X_2

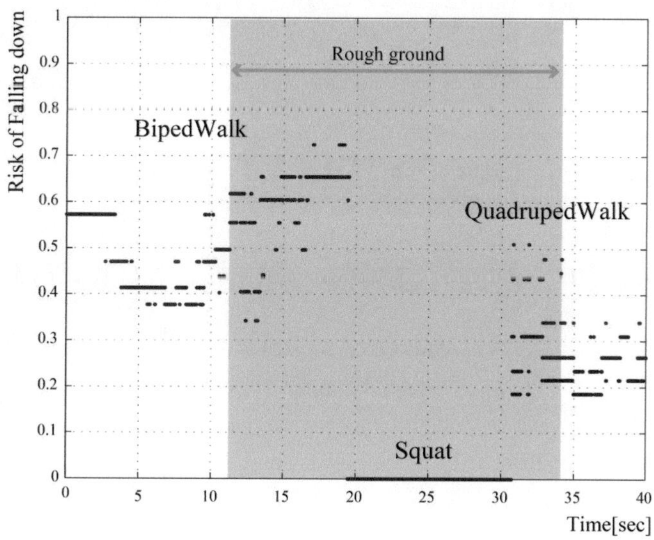

Figure 13. Risk of falling down (Experiment 1)

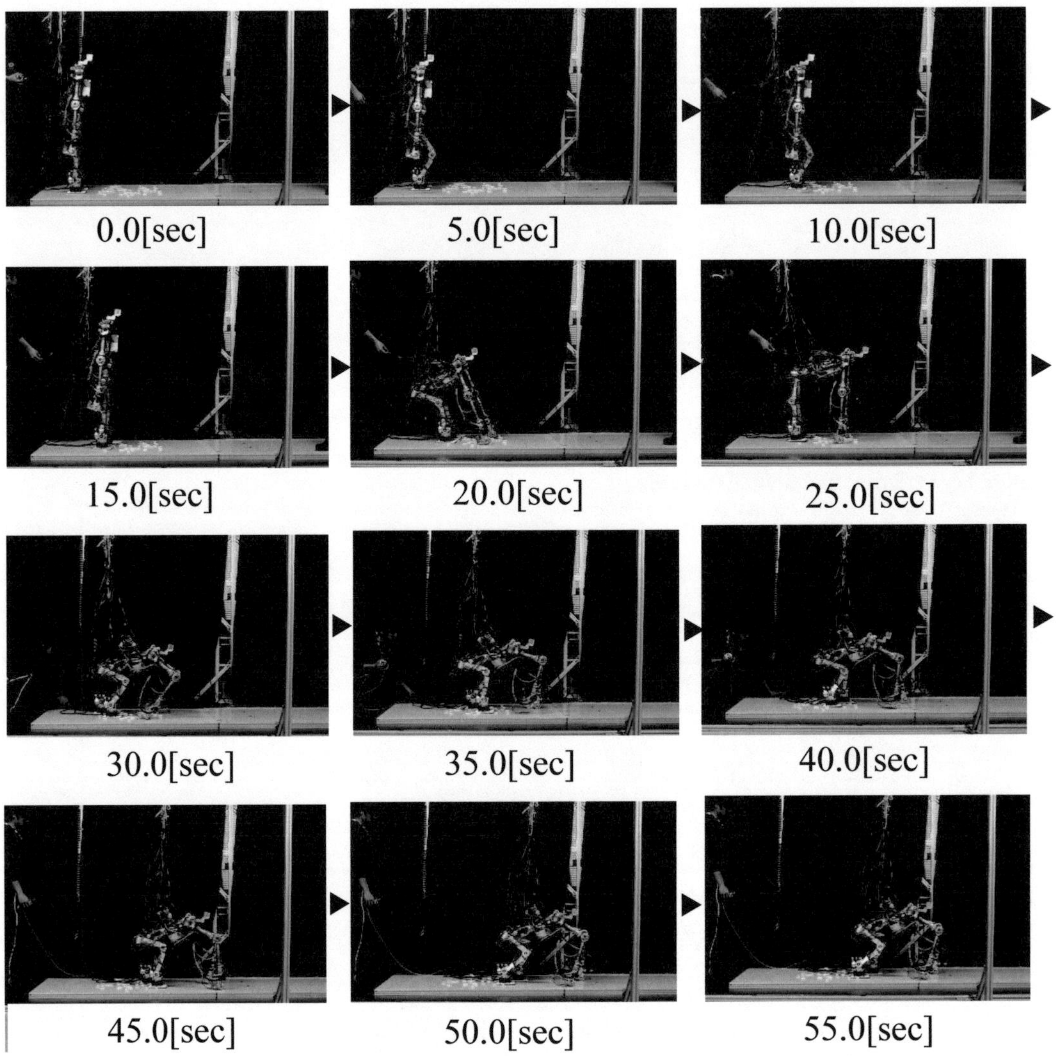

Figure 14. Snapshots of the experiment 1

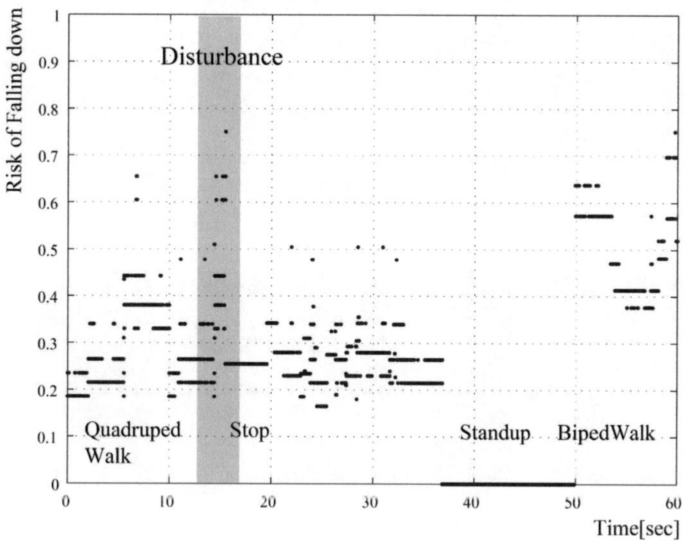

Figure 15. Risk of falling down (Experiment 2)

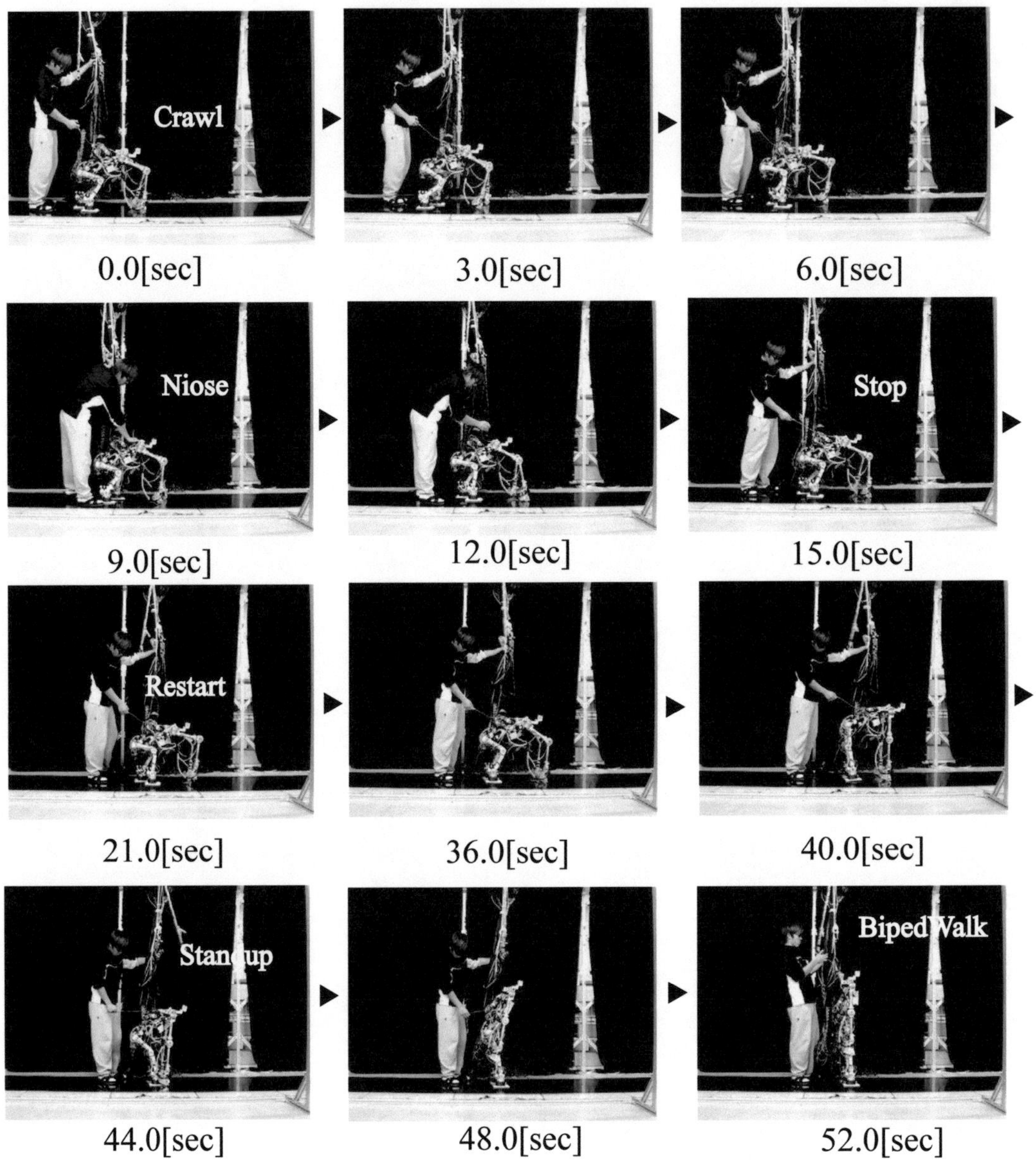

Figure 16. Snapshots of the experiment 2

5. Conclusion

This chapter firstly designed internal model composed of gait planning and stability evaluation. Next, the falling down risk is estimated by integrating stability evaluation parameters that has uncertainty using the Bayesian Network. Then we proposed the stabilization method that selects the suitable locomotion form according to the change of the falling down risk. Finally, the suitable locomotion transition is experimentally realized. Although we dealt with only biped walk and quadruped walk in this chapter, we will try to deal with other locomotion modes such as brachation and ladder climbing for transition.

Author details

Tadayoshi Aoyama
Department of Complex Systems Engineering, Hiroshima University, Japan.

Taisuke Kobayashi, Zhiguo Lu, Kosuke Sekiyama and Toshio Fukuda
Department of Micro-Nano Systems Engineering, Nagoya University, Japan.

Yasuhisa Hasegawa
Department of Intelligent Interaction Technologies, University of Tsukuba, Japan.

6. References

[1] Fukuoka, Y., Kimura, H., and Cohen A. H. (2003) Adaptive Dynamic Walking of a Quadruped Robot on Irregular Terrain based on Biological Concepts. International Journal of robotics Research. 22(3-4): 187-202.

[2] Kimura, H., Fukuoka, Y. and Cohen, A. H. (2007) Adaptive Dynamic Walking of a Quadruped Robot on natural Ground based on Biological Concepts. The International Journal of Robotics Research. 26(5): 475-490.

[3] Hirose, S. (2000) Variable Constraint Mechanism and Its Application for Design of Mobile Robots. The International Journal of Robotics Research. 19(11): 1126-1138.

[4] Wooden, D., Malchano, M., Blankespoor, K., Howardy, A., Rizzi, A. and Raibert, M. (2010) Autonomous Navigation for BigDog. In: Proceedings of the 2010 IEEE International Conference on Robotics and Automation. Anchorage, pp. 4736-4741.

[5] S. Hirose, K. Yoneda, and H. Tsukagoshi (1997) TAITAN VII : Quadruped Walking and Manipulating Robot on a Steep Slope. In: Proceedings of IEEE International Conference on Robotics and Automation. Albuquesrque, pp.494-500.

[6] Yoshioka, T., Takubo, T., Arai, T. and Inoue, K. (2008) Hybrid Locomotion of Leg-Wheel ASTERISK H. Journal of Robotics and Mechatronics. 20(3): pp. 403-412.

[7] Nishiwaki, K. and Kagami, S. (2010) Strategies for Adjusting the ZMP Reference Trajectory for Maintaining Balance in Humanoid Walking. In: Proceedings of the IEEE International Conference on Robotics and Automation. Anchorage, pp. 4230-4236.

[8] Fukuda, T., Aoyama, T., Hasegawa, Y., and Sekiyama K. (2009) Multilocomotion Robot: Novel Concept, Mechanism, and Control of Bio-inspired Robot. In: Artificial Life Models in Hardware. Springer-Verlag, pp. 65-86.

[9] Kajima, H., Doi, M., Hasegawa, Y. and Fukuda, T. (2004) A study on a brachiation controller for a multi-locomotion robot – realization of smooth, continuous brachiation. Advanced Robotics. 18(10): 1025-1038.
[10] Kajima, H., Hasegawa Y., Doi M., and Fukuda T. (2006) Energy-Based Swing-Back Control for Continuous Brachiation of a Multilocomotion Robot. International Journal of Intelligent Systems. 21(9): 1025-1038.
[11] Fukuda T., Doi M., Hasegawa Y., and Kajima H. (2006) Multi-Locomotion Control of Biped Locomotion and Brachiation Robot. In: Fast Motions in Biomechanics And Robotics: Optimization And Feedback Control. Springer-Verlag, pp. 121-145.
[12] Aoyama, T., Hasegawa, Y., Sekiyama, K. and Fukuda, T. (2009) Stabilizing and Direction Control of Efficient 3-D Biped Walking Based on PDAC. IEEE/ASME Transctions on Mechatronics. 14(6): 712-718.
[13] Aoyama T., Sekiyama K., Hasegawa Y., and Fukuda T. (2012) PDAC-based 3-D Biped Walking Adapted to Rough Terrain Environment. Journal of Robotics and Mechatronics. 24(1): 37-46.
[14] Aoyama T., Sekiyama K., Hasegawa Y., and Fukuda T. (2009) Optimal limb length ratio of quadruped robot minimizing joint torque on slopes. Applied Bionics and Biomechanics. 6(3-4): 259-268.
[15] Yoneda H., Sekiyama K., Hasegawa Y., and Fukuda T. (2009) Vertical Ladder Climbing Motion with Posture Control for Multi-Locomotion Robot. In: Proceedings of the IEEE/RSJ International Conference on Intelligent Robots and Systems. Nice, pp. 3579-3584.
[16] Aoi, S. and Tsuchiya, K. (2005) Transition from Quadrupedal to Bipedal Locomotion. In: Proceedings of the 2005 IEEE/RSJ International Conference on Intelligent Robots and Systems. Edmonton, pp.3419-3424.
[17] Aoi, S., Egi, Y., Ichikawa, A. and Tsuchiya, K. (2008) Experimental Verification of Gait Transition from Quadrupedal to Bipedal Locomotion of an Oscillator-driven Biped Robot. In: Proceedings of the 2008 IEEE/RSJ International Conference on Intelligent Robots and Systems. Nice, pp.1115-1120.
[18] Asa, K., Ishimura, K. and Wada, M. (2009) Behavior transition between biped and quadruped walking by using bifurcation. Robotics and Autonomous Systems. 57(2): 155-160.
[19] Kajita, S., Kanehiro, F., Kaneko, K., Fujiwara, K., Yokoi, K. and Hirukawa, H. (2003) Biped walking pattern generation by a simple threedimensional inverted pendulum model. Advanced Robotics. 17(2): 131-147.
[20] Sawada, H., Sekiyama, K., Kojo, M., Aoyama, T., Hasegawa, Y. and Fukuda, T., (2008) Locomotion Stabilization with Transition between Biped and Quadruped Walk based on Recognition of Slope, In: Proceedings of the 19th IEEE International Symposium on Micro-Nano Mechatronics and Human Science. Nagoya, pp.424-429.

Chapter 3

Using the Knowledge of Biomechanics in Teaching Aikido

Andrzej Mroczkowski

Additional information is available at the end of the chapter

http://dx.doi.org/10.5772/ 49955

1. Introduction

The title of the chapter refers to the research work that the author has been doing into the application of the knowledge of biomechanics in the methodology of teaching motor activities in sports disciplines involving a complex rotational movement of a human body. The author, who previously worked as a PE and physics teacher in a secondary school and an aikido instructor, now teaches biomechanics at university. Basing on his earlier work results, the author found out that some notions in mechanics are better acquired if explained using sports performance examples. According to the author, motor activities of a particular technique practised in PE and aikido classes were mastered best if their dynamics were explained to students using the principles of physics. This method also accelerated the process of understanding the rules of mechanics by the students executing a particular technique. The feelings of the students were similar - in the questionnaire made with 273 randomly chosen [1,2] secondary grammar and technical school pupils, 85% of the subjects supported explaining the rules of mechanics using sports performance examples, whereas 76% of them also supported the method of explaining techniques of the performance of certain exercises involving the rules of physics. The present chapter illustrates the experiments that verified the above-mentioned findings. The tests mainly showed the use of the knowledge of biomechanics in teaching aikido. Some of the groups of adolescents were involved in the experiments at a time interval. The first experiments carried out also checked if the effect of the knowledge of biomechanics on a shot put was increased range. The objectives of the paper are: 1. Presenting the knowledge of the biomechanics of aikido techniques. 2. Verifying whether teaching mechanics by explaining its rules using examples from aikido and various sports disciplines increases the efficiency of teaching. 3. Checking how the knowledge of biomechanics related to the rules of mechanics used in aikido techniques and shot put can improve their performance correctness. 4. Checking how a method of teaching aikido can affect the efficiency of learning aikido techniques by children.

2. Biomechanics of aikido techniques

2.1. The principles of mechanics used for executing aikido techniques

The term "aikido technique" must be precisely defined. There are multiple definitions of "sports technique" in professional literature. Bober [3] in his analysis of definitions of sports technique follows Zatsiorsky [4] who thinks that a sports technique is a term which can be described rather than defined. In the author's opinion [2], the term aikido technique refers to a way of neutralizing a specific attack and simultaneous execution of a specific motor task. Neutralizing can be made [5] by (1) locking, (2) throwing or (3) a combination of both. The neutralizing technique differs with regard to the type of attack. Aikido characteristically has a great number of techniques depending on the combination of the means of neutralization and the type of attack. Aikido is a martial art of a defensive character using the power of the attacker. In the self-defence process the following rules are applied [2,6,7]: "give in to win", "turn around if you are pushed", "move forward if you are pulled". In principle, aikido techniques should be elaborated in such a way as to make it possible for even a physically weaker person to execute them against a stronger person [8]. Simplifying the mechanical analysis, this can be confirmed by the principles of mechanics [6]. Aikido is practised mainly as a form of self-defence and it most frequently lacks sports competition. There is a clear division onto the defender executing certain techniques and the attacker against whom this technique is performed. The rules mentioned lead to a conclusion that the defender tries not to stop the attacker's move, especially with his smaller power and weight when it is impossible. If the attacker pulls with a certain force F_A (Fig.1a), then with a smaller strength of the defender F_B the result of these forces is directed at the attacker, a good solution is a sudden change of the defender's force direction into the one consistent with the force of the attacker. The resultant force being a sum of the vector values can surprise the attacker and make him lose balance. If, on the other hand, the attacker pushes with a greater force than the defender can resist, the resultant force will also have the direction of the attacker's force. In this case, a good solution would be to step out of the line of the attack.

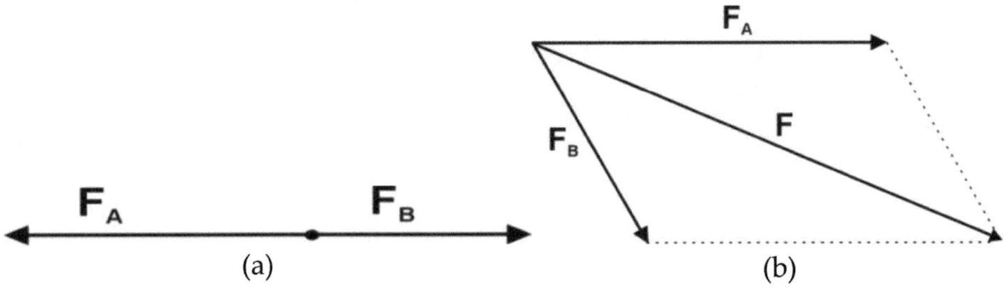

Figure 1. Generating the resultant force when the attacker (published in [6])
a) pulls the defender
b) pushes the defender

The resultant force of two convergent vectors that is then produced makes it possible to change the direction of the attacker's move (Fig.1b) The defender, by stepping out of the line of the attack changes the direction of the attacker's motion from rectilinear to curvilinear. He tries to move on the smallest possible curve. If the attacker moves in the same direction as the defender, then he additionally gains centrifugal force. If the practitioners' weights are combined by, for example, doing a grip by only one of them, then the second principle of dynamics is present here.

$$\varepsilon = \frac{M}{I} \tag{1}$$

The defender, along with decreasing the curvilinear motion radius, is decreasing the moment of inertia I of the practitioners. He tries to move in such a way as to ensure that at the end of performing the technique, the axis of motion rotation is possibly the closest to his body. When executing a certain technique, aikido practitioners are acting with certain forces, and since it is a curvilinear movement, a resultant moment of force M is produced. This moment of force, with decreasing the moment of inertia of the subjects, results in an increase in the angular acceleration ε in this motion (1). However, this analysis is of rather an approximate character. A human body is not a single solid and in a close study a biomechanical segment model of human body structure should be assumed. It is advised to apply Steiner's theorem (2) for determining the moment of inertia.

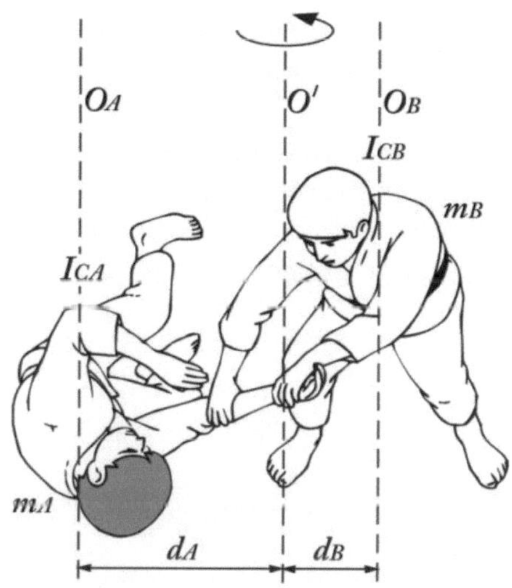

Figure 2. Analysis of distribution of the practitioners' masses around the common axis of rotation O′ in the final stage of the technique

$$\begin{gathered} I' = I_c + md^2 \\ I' = I_A' + I_B' \quad I_A' = I_{CA} + m_A d_A^{\ 2} \quad I_B' = I_{CB} + m_B d_B^{\ 2} \end{gathered} \tag{2}$$

The moment of inertia I′ is a sum of the moments of inertia of the subjects, namely the attacker I_A' and the defender I_B'. For this purpose, their central moments of inertia I_{CA} and I_{CB}, as well as the distances d_A and d_B from their centres of mass m_A and m_B (Fig. 2), must be determined. When the radius on which the attacker is moving decreases, the moment of inertia of the position of the practitioners' bodies gets smaller and their angular velocity rises. If we neglect the motion resistance, we can talk about the principle of the conservation of the moment of momentum

$$I\omega = \text{const} \tag{3}$$

The subjects behave similarly to figure skaters when performing a pirouette. In this figure they move their lower limbs close to the axis of rotation and by doing this they increase their angular velocity ω. In some moment of the motion, the centre of mass of the defender should be at the closest possible distance from the axis of rotation of the performers' bodies' arrangement. His hands should be as close to this axis as possible. This leads to a decrease in the moment of inertia of the performers and to an increase in the value of the centrifugal force F acting on the attacker.

$$F = \frac{mV^2}{r} \tag{4}$$

As the formula (4) shows, the centrifugal force gets bigger when the velocity the attacker attacks with goes up, his weight increases and the radius he follows gets smaller. The behaviour of the attacker can be compared with the behaviour of a car on a road bend. The sharper the bend and the smaller the radius r, the bigger the car speed V and the bigger force acting on the car, increasing the risk of falling off the track. The behaviour of the defender resembles the motion of a spinning top (Fig. 3). The external force acting on the spinner cannot disturb its rotational movement. The defender makes a move causing the attacker and not the defender to gain the centrifugal force. Therefore, he performs stepping out of the line of the attack in such a way so as the whole motion is done around the axis of rotation going most desirably through his body.

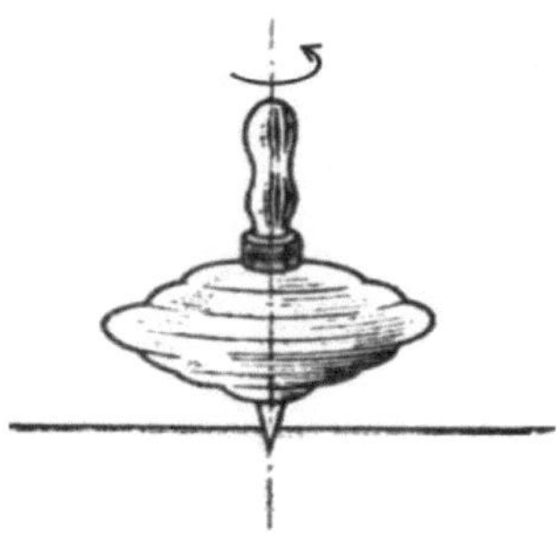

Figure 3. Spinning top (published in [6])

The centrifugal force may allow neutralizing the attack. However, it is usually too small to knock the attacker over. In order to do a throw, the defender uses his weight, which when adequately transferred may exceed the centrifugal force gained by the attacker [6]. Therefore, in a great number of techniques, the defender suddenly lowers his centre of mass in order to increase the technique dynamics. In the final stage of the throw, provided the attacker's body is inclined enough, in order not to meet his unexpected counter-punch, the defender is even likely to do a one leg jump. When performed correctly it facilitates the shift of the defender's weight down. Then all his P=mg is used in the technique. Most frequently the force enabling an aikido throw is a result of the centrifugal force of the attacker and the weight of the defender (Fig. 4). A good example illustrating the conduct of the defender would be a spinning top that apart from a rotational motion would do an up movement, such as jumps. An approximate formula for a resultant force producing a throw can be determined as follows:

$$F = \sqrt{\left(\frac{m_2 V^2}{r} + \left(m_1 g\right)^2\right)} \tag{5}$$

Figure 4. Resultant force acting on the attacker in the final stage of the technique (published in [6])
m_1 – the defender's mass
m_2 – the attacker's mass

The figure does not show all the vectors of the force, which can be obtained by means of, for example, pelvis turns, characteristic for aikido performed by the defender along with the body turns around in a horizontal plane, or by means of a force coming from the use of particular muscles of the attacker. Many authors explain the dynamics of the defence techniques (in martial arts) quoting the principles of biomechanics [6-11]. Of special interest here are the lectures of Jigoro Kano explicating the "give in to win" principle. The father of judo was familiar with the biomechanical aspects of judo. The interplay of centrifugal and centripetal forces or movements resembling a spinning top involved in the execution of aikido techniques was understood by the son of the aikido founder Kishomaru Ueshiba [12]. Koichi Tohei [13], the only one who was awarded by the founder of aikido when he was still alive with the 10th Dan, claimed that the secret of Morihei Ueschiba was his ability to relax when executing the techniques which was also due to a low position of the centre of mass. However, it is not a complete loosening of muscles, but rather straining the muscles which, for example, causes a child to have such a power that another person cannot snatch the child's favourite toy out of his hands. This ability of relaxing or loosening referred to above, is related to the concept of "ki". The idea of "ki" has rather a wide meaning in the Japanese culture and it is difficult to translate it into a European meaning. Generally, it is understood as life energy possessed by every man. However, an explicit explanation of "ki" in terms of mechanics is at the current level of research limited and thus goes beyond the subject matter of this presentation. The breathing techniques generally applied in some aikido schools understood as developing "ki" make it possible to master the ability to relax/loosen when doing a throw. In terms of throw dynamics in aikido it gives a greater possibility to shift the force coming from the weight of the defender P=mg.

3. Biomechanical analysis of aikido techniques executed by the disabled

It is obvious that in combat sports competition between a disabled person, for example, missing one limb, and a fit person is practically impossible. Lack of full power in one of the limbs gives a significantly smaller attack power and its potentials. In the case of a necessary self-defence, a disabled person, by using the power of the attacker, has a chance to execute some aikido techniques. Below, the author gives a mechanical analysis of some aikido techniques performed by disabled persons [14]. Together with the pictures you will find figures presenting force vectors in two planes. It would be more precise to show the vectors acting in a 3D plane, but this would not be readable in this paper. For convenience, Fig. 5 illustrates vectors in a horizontal plane and in Fig. 6 they are shown in a vertical plane.

3.1. Execution of aikido techniques by people with a dysfunction of an upper limb

The technique (Fig. 5) illustrates a defence against an attack made by hand in a circle throw at the head level. The defender steps out of the line of the attack, grasping the attacker's hand. The attacker works with force F_A, the defender with force F_B, their vector sum gives vector F (Fig. 5). In the next stage of the technique under analysis, the defender lowers his centre of mass. In this way he supplements the force configuration with force G that comes mainly from the weight of the defender (Fig. 6). This force is of a great importance in the dynamics of aikido techniques, provided it is transferred at the right moment of the technique [6].

Figure 5. Defence against an arm attack following a circle at the head level – analysis of forces in a horizontal plane (published in [14])

In the vertical plane, force F and the force G give a resultant F_W. When executing a technique the defender does a turn around in order to shorten the radius followed by the attacker. The attacker is acted on with the moment of force equal to the product of the F_w resultant and the radius on which the attacker is moving.

The moment of the force produced causes the attacker to lean forward and lose his balance.

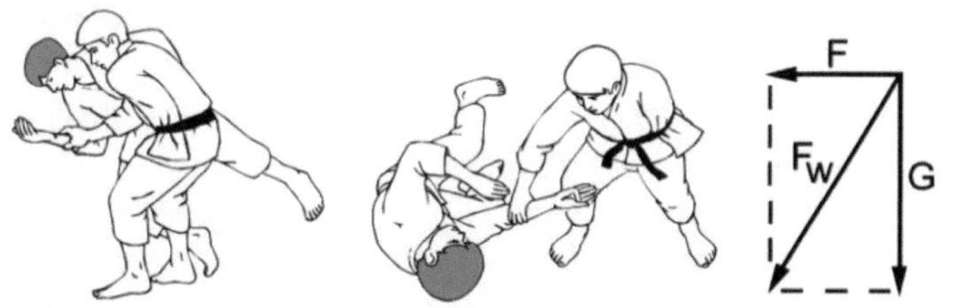

Figure 6. Defence against an arm attack following a circle at the head level – analysis of forces in a vertical plane (published in [14])

The above-mentioned aikido technique can be performed by a disabled person using only one upper limb. Force G is the most important as far as the dynamics of this technique execution is concerned. For applying this force, only one upper limb is needed, because only one point of the force application is enough. The application point of force G is supposed to be like "the eye of a cyclone". It is a central place where movement is the smallest. Force F_b changes the direction of the attacker's move (its value does not have to be big). Only one hand is needed for this change of direction, whereas the other can, for example, shield the body. With adequate speed of the defender, this technique can be executed with one hand neglecting shielding of the body. As previously mentioned, the movement of the defender in aikido often resembles the motion of a spinning top [Fig. 3], which apart from rotating also executes an upward movement. At the end of the technique the arms are in most cases placed close to the axis of the body rotation.

Figure 7. A form of a defence - grasping the attacker's arm with both hands (published in [14])

Aikido comprises many techniques using a movement of only one upper limb. These techniques are executed following the same principles of mechanics as shown in Figures 5 and 6. There are many possibilities of mechanical motion versions, thus, their description can be too much expanded upon. Figure 7 presents an example of a more complex technique. The attacker grasps the defender's hand with his both hands. As a result of stepping out of the line of the attack and shifting the body mass down along with the movement of the arm, the first resultant force F_{w1} is obtained. This force, as in Figure 7, is a result of forces F and G_1 coming from the mass of the defender. In terms of the technique, force F_{w1} is used mainly for a correct leaning of the attacker and for giving speed. Then, after moving the centre of mass up and down again, along with the movement of the hand, force G_2 is produced that comes mainly from the mass of the defender. Roughly speaking, the composition of the forces G_2 and F_{w1} gives a resultant F_{w2} causing the attacker's fall. It is quite easy to select from the aikido repertoire the techniques for which only one upper limb is used. Thus, these techniques can be used by disabled persons who have only one efficient arm.

3.2. Aikido techniques for people with a dysfunction of a lower limb

Some aikido techniques involve the *hanmi handachi waza* position for their execution, with the attacker in a standing position and the defender in a kneeling position. This practice dates back to ancient Japanese rituals, where people used to have meals and relax in kneeling positions [5]. A samurai in this position was prepared to defend himself against a sudden attack of his opponent by means of certain defence techniques that he had mastered. Many aikido masters have claimed that executing aikido techniques in a kneeling position particularly influenced their execution in a standing position. Aikido techniques performed in a kneeling position can be executed by people with certain dysfunctions in lower limbs, for example, as a result of limb amputation below the knee. It is crucial that the defender maintains the point of support on his knees. Figure 8 illustrates an example of a technique executed by a kneeling defender. The technique presented is a defence against the *shomen uchi* attack with an open hand strike downward in a vertical plane. The biomechanical analysis is based on the same rules of mechanics as the techniques referred to above. Figure 8 shows the forces of F and G_1. Similarly to the situation in Figure 7, stepping out of the line of the attack produces the centrifugal force F, whereas his correct hand movement (ended downward) adds force G_1. As a result force F_{w1} is generated. The defender, by grasping the attacker from behind with his second hand, causes that the second resultant force is produced (as in the situation illustrated in Figure 7), F_{w2}. Unfortunately, it was not possible to mark this force in Figure 8. In kneeling positions stepping out of the line of the attack is rather limited due to a smaller speed of movement in comparison with the movements of standing practitioners. Some of the techniques executed in this position require mastering a special method of moving around, namely *shikko*. This method of moving around can be adequately adapted depending on the degree of motor dysfunction of a disabled person. It has been shown that using this method of moving around along with the selected aikido exercises can have a beneficial effect on the health of children with pelvis placement disorders in frontal plane, as well as with a lower degree scoliosis [15-18].

Figure 8. Defence in the *hanmi handachi* position against an open hand attack in a vertical plane (published in [14])

3.3. Execution of aikido techniques by people in a wheelchair

The possibility of making a quick move when sitting in a wheelchair is limited. As far as aikido techniques are concerned, such people can only do a leverage. This mainly means locking wrist joints. An example of this technique is shown in Figure 9. The defender is trying to grasp the hand of the attacker, then he performs a *kote gaeshi* leverage which means wrist twisting. The possibilities of using aikido exercises for the disabled have been confirmed by Rugloni [19].

Figure 9. Defence against a fist attack by doing *kote gaeshi* (published in [14])

4. Materials, methods and experiment results

4.1. Experiment I

4.1.1. Materials and methods

The experiment started in September 2000 and involved 200 pupils (15 16 year olds) attending six secondary school first classes (secondary grammar school and secondary technical school) from a city in Poland [1,2]. The classes were randomly chosen. Mechanics is a part of the first physics class in the course. Physics was taught in four classes (one in the secondary technical school designated for the purpose of the experiment as F and three in the secondary grammar school designated as C, D, E) with a total of 137 students being taught by explaining the principles of mechanics with examples taken from sports practice and by simultaneous explaining the execution of particular motor activities of the given sports techniques, viewing their dynamics from the perspective of physics. Physics in the other two control classes (one in the secondary grammar school designated as A and the other in the secondary technical school designated as B) with 63 students in total was taught using a traditional method. The physics course in classes A, C and D contained one hour of teaching and in classes B, E and F two hours of teaching. The classes A, B, C, D and E were coeducational (mixed), whereas F was boys only. Both groups used the same class books recommended by the physics curriculum for first class technical and grammar school students, however, in the test group many task examples had slightly different contents. This meant that, for example, a task instruction "a body moves down an inclined plane" was replaced with "the ski jumper Adam Małysz skis down the ski-jump". A part of such examples was taken from the book references on sports biomechanics. Before the experiments started at the beginning of the school year 2000/2001, the experimenters checked the subjects' knowledge of mechanics that they had acquired in elementary school and the degree of their understanding of the dynamics of the randomly selected sports techniques in terms of mechanics. The test in mechanics aimed at checking the students' understanding of the principles of mechanics and not their total knowledge of this subject, for example, in a form of a definition included in a particular unit of the physics course. A similar test was carried out at the end of the school year, after the students had completed the mechanics course. Both tests were unannounced and when doing them students could use the formulas covered by physics reference tables or written down by a teacher on the blackboard. The experiment also involved researching the influence of teaching mechanics on the progress achieved in sports performance exemplified by putting a 5 kg shot. 25 boys from the second class of a secondary grammar school were randomly selected for the experiment (16-17 year olds) from the same area as the subjects from the first classes. Firstly, the boys' results in shot putting facing sideways were checked. Then, they were taught the putting mechanics. The experimenters explained what principles of mechanics affected the range and on what motor activities attention should be focused in order to improve to shot put. They also learned why, in terms of mechanics, the shot putting technique facing sideways results in a smaller range than facing backwards or why it is smaller than in a rotational technique. Basing on book references, the students acquired theoretical knowledge of the shot putting technique and its mechanics. In the next PE classes they were taught shot putting facing backwards with the stress put on understanding the technique by explaining it with the principles of mechanics. After three weeks students did a written test in shot put mechanics, which merely checked their understanding of the rules, and then the results of shot putting facing backwards achieved by students of the test group were checked. The tests were carried out in comparison to the initial tests, weather and time conditions. The results were analysed by the analysis of variance. In the first test the significance of differences was established by means of the 't' Duncan test, whereas in the second test a relation between an increase in the shot put range and the degree of mastering the putting mechanics was established by means of the analysis of regression with $p<0.05$.

4.1.2. Results

4.1.2.1. Effect of teaching physics based on sports practice examples on the degree of acquiring the knowledge of mechanics required by the curriculum

The average score in the written test carried out at the beginning of the school year was rather poor (2,0) and no significant differences among students from various classes under analysis were spotted. After completing the course in mechanics required by the curriculum of the secondary grammar school first class, the experimental group that was taught physics with examples from sports practice achieved a significantly higher test score average than the control group that was taught in a traditional way (Fig. 10). The score scale was from 1 to 5, where the worst score was 1 and the best was 5.

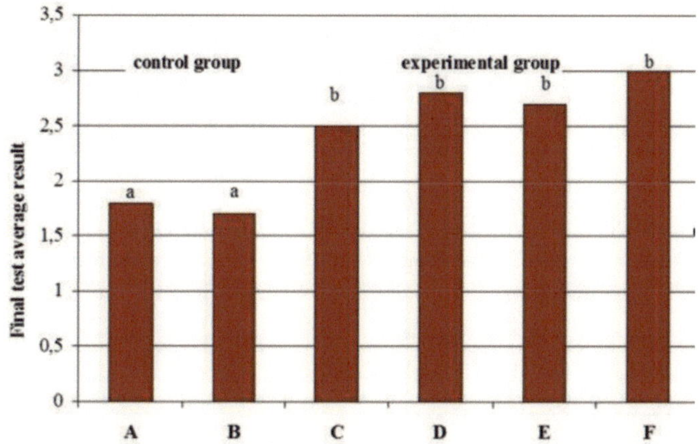

Figure 10. The average score in the test - in mechanics reached by students taught physics in a traditional manner (control group) or basing on the sports practice examples (experimental group). The averages marked with different letters differ from one another according to the 't' Duncan test with $p<0.05$ (published in [1])

In the experimental group, students' skills in explaining the correctness of executing certain motions in the selected sports techniques with the principles of mechanics was also checked. It was found that at the beginning only 22% of the subjects possessed these skills, whereas at the end of the experiment this value grew to 78%.

4.1.2.2. Effect of the intensity in teaching physics on the degree of acquiring the knowledge of mechanics required by the curriculum

Contrary to the expectations, increasing the number of physics classes from one hour to two hours a week did not significantly improve students' understanding of the principles of mechanics, both in the control as well as in the experimental group (Fig. 11).

4.1.2.3. Effect of teaching mechanics on the range of shot put

A clear relation between sports performance results and the degree of mastering the theory was found in the group of pupils to which the shot put technique was explained with the laws of mechanics (Fig. 12).

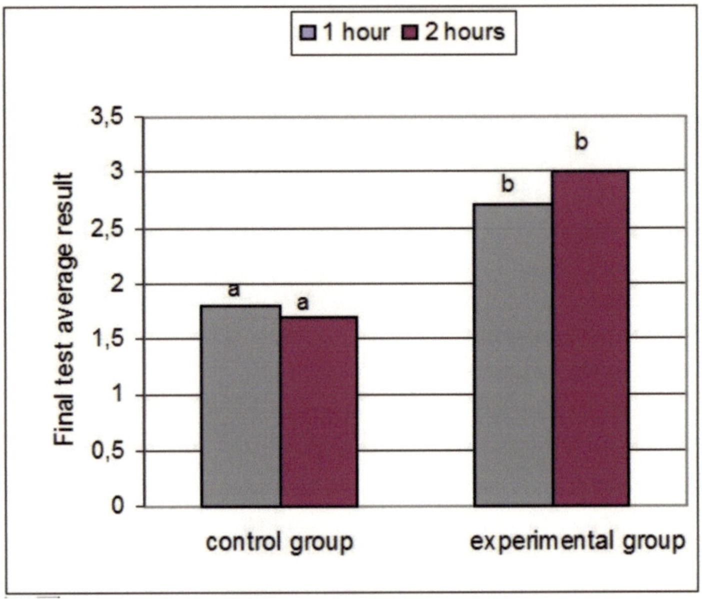

Figure 11. Effect of the method of teaching physics and its intensity on the degree of acquiring the knowledge of mechanics by pupils. The averages marked with different letters vary from one another in accordance with the 't' Duncan test with $p < 0.05$ (published in [1])

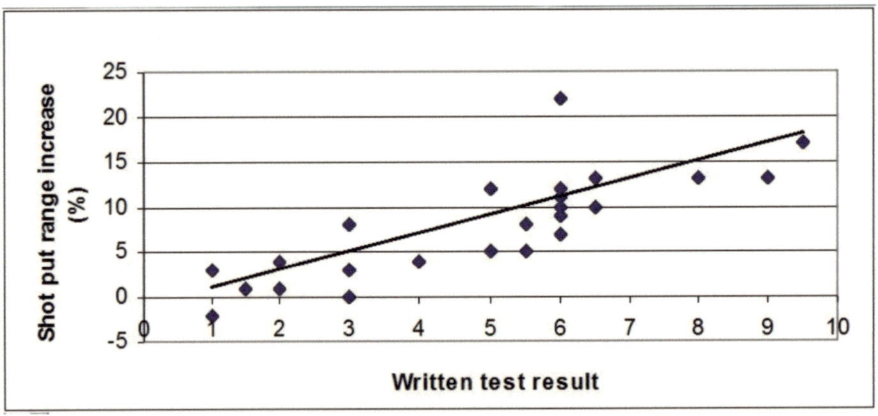

Figure 12. A relation between the shot put range increase and the results in the written test in mechanics (y=2,26 + 3,24x, r=0,83) (published in [1])

The score scale for the written test in mechanics was from 0 to 10. The putting range increase was expressed as a relation between the shot put range increase (Δy) after completing the course of this put mechanics and the range (y) reached at the beginning of the experiment. During the test 23 students out of 25 performing the test tasks improved their sports result, one got the same result and in one case the participant's result was slightly worse. A strong relation between relative shot put range increase and results in the written test in mechanics was found. This is confirmed by the high value of the correlation factor r=0,83 obtained. After doing the theoretical test and after completing the practical test in shot putting, the students reported in writing on what elements of their technique facing sideways needed improving in terms of mechanics. It turned out that 87% of students were able to state what they should improve.

4.2. Experiment II

4.2.1. Materials and methods

The samples for the study consisted of two groups of secondary school students from a city in Poland [2]. All subjects had participated in similar studies in the school year 2000/2001 [1]. The two study groups were in fact two school classes E and F. They were 17 and 18 years old. Group E (control group) comprised 33 students and group F (experimental group) included 27 students. In the first semester of the school year 2001/2002 group F (second secondary school grade) practised aikido one hour a week in an additional PE class. Solid-state mechanics is part of the physics curriculum in the third grade of secondary school, after students acquire sufficient mathematical skills in their lower grades. Before the commencement of the mechanics course in the school year 2002/2003 students had to pass a physics revision test covering material from the first grade, i.e., translational mechanics (test 1). Solid-state mechanics was taught in class E in the traditional way, whereas students from class F used their experiences and examples from aikido acquired during the additional PE class. This mechanical movement was also compared with other sports techniques used in diving, sports gymnastics, dancing, figure skating, etc. Both groups then took a final test to assess their learning outcomes. Both tests (1 and 2) were surprise tests and were carried out sometime after the last class to ensure assessment of students' understanding of the mechanics principles, and not merely their memorizing skills. During the additional PE hour the F class practised aikido. The teacher explained the dynamics of aikido techniques using the principles of biomechanics. He also explained what biomechanics dealt with and what relation it had with the mechanics taught at school. The students were told that biomechanics applies the rules of mechanics in analysing human movements. In describing this movement the human body is divided into 14 parts treated as solids [20]. The classes focused mostly on the solid-state rotational mechanics. The participants were informed about the advantages of reduction of the radius of performed movements, the distance between the arms and the axis of rotation, or of lowering or optimal shifting of the body's centre of gravity. The principles of aikido, such as "yield to win", "turn around if you're pushed" or "move forward if you're pulled", were explained to the students using the law of momentum conservation, second law of motion for angular motion, centrifugal force and composition of resultant forces and moments of force [6, 7]. The students' knowledge of biomechanics for aikido was tested in essay form. This test checked students' understanding of executing motor activities involved in the techniques performed with the rules of mechanics. After writing their essays, the students got acquainted with four selected aikido techniques over a one month period. The selected techniques in the form of their performance made it possible for a defender to use the centrifugal force acting on the attacker, as well as their mass as in Figures 6 and 7. Then the precision of performance of each technique was assessed using a ten-point scale. Each subject could score up to 40 points. The method of evaluation of aikido technique performance was taken from the Koichi Toheia aikido school. As sports rivalry does not in principle apply in aikido, practitioners are evaluated on their performance of *taigi*, i.e., sets of techniques executed in response to a particular attack. Both the precision and speed of movements are evaluated. During the study only the precision of movement sequences performed by a practitioner was assessed. The precision criteria included an appropriate reduction of the radius of motion, assuming proper body posture, arms movement and shifting of the body's centre of gravity during a given movement. Before learning the selected techniques, the students taking part in the test had already acquired basic aikido skills, i.e., safe falls, body turns and rotations. The assumed method of instruction was "from the general to the specific" [2,7,21]. This method stresses first of all synthetic teaching by explaining the general rules of technique execution which is only then followed by an analytical study of particular movement sequences in a given technique.

4.2.2. Results

The effects of the mean results of the written tests in both groups were assessed using the student t-test for independent variables. No statistically significant differences were observed between group E and group F in test 1, but not in test 2 ($p < 0.05$) (Fig. 13), where group F attained a much higher mean result than group E. The obtained student t-test results were confirmed by results of the non-parametric U-Mann Whitney test.

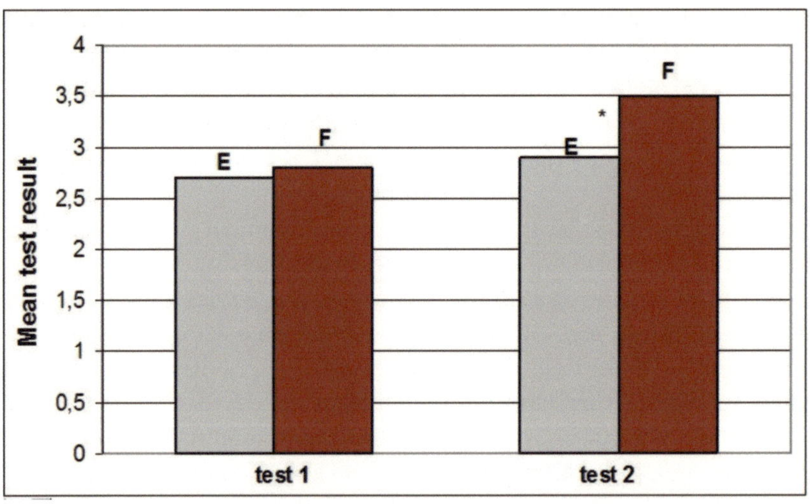

Figure 13. The mean results of mechanics tests taken by students from group E (taught in the conventional way) (n = 33) and from group F (taught using examples from aikido and other sports) (n = 27) (published in [2])

The analysis of regression was used to examine the correlation between the correctness of the biomechanics test answers and the precision of performance of aikido techniques. A strong correlation was noted between the two at $r = 0.9$.

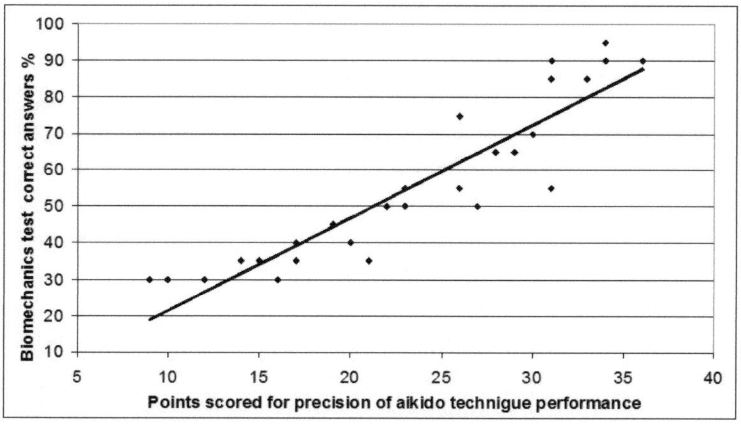

Figure 14. Correlation between results of the biomechanics tests and points for precision of performance of aikido techniques. Equation of the regression line: $y = -4.10 + 2.56\,x$, $r = 0.9$ (published in [2])

4.3. Experiment III

4.3.1. Materials and methods

The research involved fourth form pupils at a primary school from a town in Poland aged 10-11, divided into two 25-strong training groups (A and B) [7]. The pupils were selected at random. The two groups thus formed were representative enough for the whole of the school's pupil population. The pupils at the school came from one district, so the groups were homogenous because the pupil qualification to separate into groups was based on random selection (selection does not considered special skills, intelligence, material or social status, achievements of previous education years etc.). The experiment was performed during the first term of the school year of 2001/2002. Similar research methodology used to teach aikido to secondary school students was used to teach primary school students [2]. However, due the difference in age of the participants, the methods were slightly modified. The subjects practised aikido at extra-curricular PE classes. Having mastered safe falling - necessary for the practice of aikido - the pupils were taught four selected aikido techniques over a period of one month. Different methods of teaching were applied for particular experimental groups. A were taught using the 'general-to-specific' method [7,21], whereas B were taught using the 'specific-to-general' method, with the analytical approach prevailing. Successive movements of a particular technique were taught. Only after the children mastered the particular movements, were they taught to put them together in order to perform a particular technique. The children learned the movements by following closely the instructor's movements. Group A were taught using the 'general-to-specific' method, with the prevalence of the synthetic method: for instance, teaching a particular technique's proper movements involved decoding the motor abilities the children already possessed. A good idea is, for instance, to enact a "crowded underground station". The trainees occupy a small space in a room thus making an artificial crowd. At a signal they start moving fast. It can be seen that some of them, in order to avoid a collision with others, in a natural way make turns and revolutions. It may happen that they have already mastered the proper way of turning and revolving - and they only need to retrieve it. During the instruction, the instructor first tried to find out which turns and revolutions the children had already mastered. Next, he showed how to use them in a particular aikido technique. Only the lacking abilities were taught from scratch. Group A was taught aikido, drawing on the knowledge of biomechanics [6]. This was a completely new thing to the children since physics is not included in the primary school curriculum. The techniques were explained by, for example, quoting the second principle of dynamics of rotary motion, the principle of conservation of angular momentum or the factors influencing the value of the centrifugal force. Because the children at this age have no mathematical knowledge that would allow them to understand the principles in the form of formulae, the particular laws of physics were taught by means of experiment presentations, without the use of scientific terminology. Instead, children's language was the one of instruction. The presentations included observing a top spinning. The children played 'top fights' on special boards, by making the tops move in such a way so that that the top's motion imitated some aikido techniques. However, they were also made aware that those techniques also involve vertical motion - resembling a top which suddenly rises or lowers its position while spinning. The children were told that this additional upward movement enables the practitioner to take advantage of his mass while performing a defensive technique. Next, the children were shown a sample technique involving this kind of motion. The children were also allowed to try and execute the technique - imitating the movements of a top. Referring to biomechanics, the instructor explained aikido teaching principles [6,21], such as 'turn around if you are being pushed', 'go if you are being pulled', or 'give in to win'.

After experimental interpretation of the principles of mechanics present in aikido techniques, Group A children were tested for their understanding. By asking questions and using questionnaires, the children's understanding of the movements in aikido techniques was assessed basing on the experimental interpretation of the rules of mechanics covered. A sample question went: 'Using knowledge of biomechanics, how do you move a particular body part in order to execute a particular aikido technique most effectively?' After questionnaire-based oral testing in Group A, both groups were taught four aikido techniques. This research method was also used by this author when carrying out research on secondary school students [2]. After one month, the children were tested for their execution of the techniques. Similarly to the situation of secondary school students, the assessment focused on the effectiveness of execution of a particular sequence of movements, regardless of the tempo which was to be slowed down. The tempo of the attacker was adjusted so that the defender could execute all the successive movements of a particular technique. The correctness of the execution was assessed by this author using a 1-10 point score scale. The maximum score was 40 points. To analyse the findings, mathematical statistics methods were used, including analysis of regression and student t-test for the independent variables. Analysis of regression was used to find out about the correlation between the correctness of answers to questions concerning biomechanics and the correctness of aikido technique execution. The relation between the scores for the execution of aikido techniques and the type of group was examined using student t-test for the independent variables.

4.3.2. Results

Group A achieved a much higher arithmetic mean than group B (Fig. 15) as far as aikido technique execution is concerned. It was also found out that the results achieved for both groups differ statistically ($p<0.05$). A high statistical significant correlation coefficient ($r=0.75$ for $p<0.001$) was found between the correctness of the answers to questions concerning biomechanics (Fig. 16). This is confirmed by a high value of the correlation coefficient r=0,75.

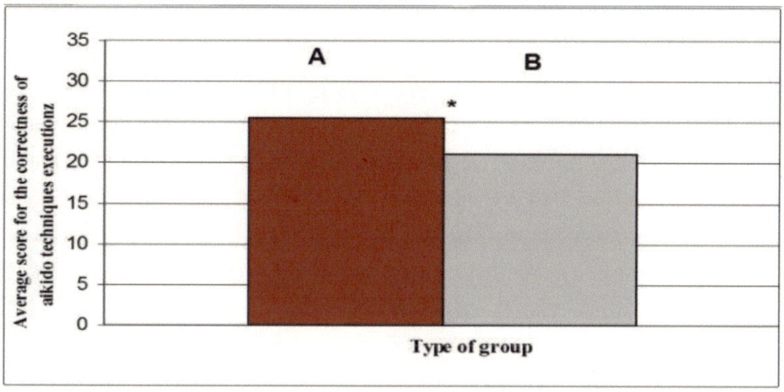

Figure 15. Average scores of children in both experimental groups (A and B) for the execution of aikido techniques (*p<0.05) (published in [7])

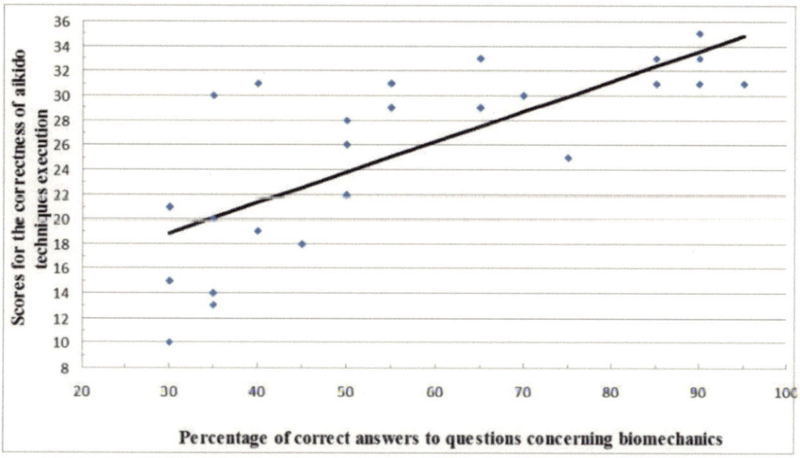

Figure 16. Correlation between the scores for the execution of aikido techniques and the correctness of answers to questions on biomechanics in Group A (n=25). Simple regression equation: $y=11.538+0.245x$, $r=0.75$ (published in [7])

5. Summary

5.1. Mastering the knowledge of biomechanics

The results of Experiment I show rather poor average results achieved by the participants in the mechanics test, especially in the first test that covered the elementary school course of this subject. This may be explained by the fact that it was not an easy test, because it checked understanding of the principles of mechanics. Since it was a surprise test it checked students' durable knowledge rather than the information they could remember from lesson to lesson. Such knowledge is based on understanding, most conveniently backed with examples that stay in one's memory, like sports examples. A student, remembering an interesting case will always refer it to the principle of mechanics related to that example. It is interesting that in a control group the average score in the final test lowered as compared with the results of the test the students did before the commencement of the experiment, whereas in the experimental group the final test results were better than the results of the first test. However, the comparison of the results of both tests should be treated as some kind of simplification. First of all, the initial test checked pupils' knowledge of mechanics on an elementary school level, whereas the final test checked this knowledge acquired by first class secondary school pupils. Nevertheless, in both tests understanding the mechanics and not learning rules was focused upon. Thus, the fact that the control group classes got lower results in the final test than in the first one, does not mean that they did not make any progress. It is certain, though, that pupils in the experimental group made much greater progress in understanding the principles of mechanics in comparison with the control group. The boys' class achieved the highest average score, however, the difference between their results and the results of mixed classes in the experimental group was not statistically significant. This leads to the conclusions that examples based on sports performance are equally attractive to boys as to girls. It is intriguing though that increasing the number of classes in physics a week did not affect the average score a particular class got in the test. Thus, the conclusion is that for understanding the rules of mechanics the subjects needed only one hour of physics a week. A significant factor resulting in increasing the test result was definitely a more attractive method of teaching. A great difficulty in teaching physics at school is the fact that pupils' knowledge of mathematics used for describing the laws of physics is not satisfactory. Usually, owing to a small number of classes a week, physical experiments are done by a teacher in demonstration form and pupils cannot practise doing them by themselves. Sometimes, a pupil is required to learn definitions by heart instead of focusing on understanding them. All this leads to a situation in which it is hard to achieve the most important goal, that is to make pupils interested in the school subject they are being taught. The test results show that pupil learning of the principles of mechanics is facilitated by using sports examples, which generates their increased interest. A pupil performs in a PE class a mechanical movement. The only problem is to make him aware of the fact that it is a perfect experimental base which can be referred to during classes in physics. There was no statistically significant difference between the two groups of students (E and F) participating in Experiment II in terms of their results of mechanics tests in the school year 2000/ 2001(experiment I), their average scores were similar. This was confirmed by the results of the written test taken by the students before the start of the solid-state mechanics course (test 1). It shows that both groups had similar conditions for learning the mechanics material. A new method of teaching solid-state mechanics significantly affected the differences between the results of solid-state mechanics learning attained by both groups of students. In a traditional physics class a teacher demonstrates an experiment using the equipment of the physics study. In the author's opinion the main factor affecting the results in group F was students' participation in aikido classes, during which aikido techniques were explained using the principles of mechanics. The students' participation in aikido classes could facilitate their understanding of mechanics principles, following the educational rule of affecting as many of the students' senses as possible [2,7]. The PE classes as well as other forms of extra-school physical activity allowed more active contribution of other senses – including somatic ones – to the learning process, than hearing and sight only. The obtained results correspond to the results obtained in experiment I. Learning the laws of mechanics by school students could be greatly facilitated by the use of examples from sport and other extracurricular physical activities [1].The experiment III differs from the previous study in the way the knowledge of biomechanics was conveyed. The previous study made use of the mathematical formulae describing the principles of mechanics which the students had come to know in physics classes. This was all but impossible in the present research, since the pupils had not had physics at school yet. The analysis of the questionnaire used in the experiment leads to the conclusion that the methods used for teaching particular aikido techniques facilitate children's understanding of the laws of physics describing rotary and translatory motion. Children may learn to predict the results of sudden changes of the direction of their movement, e.g., turning suddenly while running straight. In this way they may learn about the forces involved in such changes and about ways of using those forces for a particular purpose.

5.2. Effect of mastering the knowledge of biomechanics on technique execution

The results obtained in the shot put (experiment I) lead to the conclusion that the increase in the results was caused by turning from the shot put facing sideways to the technique of facing backwards. It should be excluded that the increase in the result was affected by an increase in the pupils' fitness, because in such a short period of time (3 weeks) the group participating in the test could not improve it to a great extent. The range increases achieved were caused by changing the putting technique. The results in experiment II and III show that the aikido techniques were mastered best by those who were better at understanding the principles of mechanics involved in the techniques, from which it follows that it is justified to explain how to perform a particular technique quoting biomechanics. The concept of motor teaching which relies on the awareness and understanding of the task while learning a specific move, proposed by Bober and Zawadzki [20], appears to be a valid one, too, with regard to the teaching of aikido to children. Such a teaching method not only facilitates the repetition of already mastered moves, but also helps invent new ones according to the needs. In experiment III the aikido classes employing the "general-to-specific" method resulted in a better mastering of aikido techniques than the ones based on the "specific-to-general" method – as is confirmed by Group A's higher average score for the execution of aikido techniques, compared with Group B. Thus, the synthetic way of teaching the execution of the techniques gave better results than the analytic teaching of successive moves involved in a particular technique [2,7,21]. With Group A, an experimental interpretation of the laws of mechanics, describing rotary and translatory motion of the human body, was used. In this case the interpretation relied mainly on using the partner to explain the laws of physics governing translatory and rotary motion. Performing a particular aikido technique with a partner gives one an opportunity to feel the effects of the force exerted on the partner and vice versa. Such a possibility to experiment has been noted by Kalina [22]. According to him, what X aikido student experiences from Y aikido student, will in a moment be Y's experience from X . Such participation in mechanical motion may help one understand mechanics - in compliance with a pedagogical permanence principle of 'stimulating most of a learner's senses' [2,7]. The instruction based on the use of the knowledge of physics in martial arts may not only be used with techniques, but also to teach safe falling. Simple falling experiments with a notebook or a cushion [7] may explain the principles of safe falls in judo or aikido - by showing how a falls' mechanical energy can be reduced by increasing the body surface that hits the ground [23, 24]. A show experiment may improve the child's understanding of the importance of positioning oneself properly while falling. The method adopted for the needs of this experiment did not require the children to display motor abilities at a certain level because of the form of instruction, a presentation of the execution of techniques. This researcher's assessment method applied to a slowed-down execution of aikido techniques, which was necessitated by the need to limit the effect of the subjects' motor fitness upon the effectiveness of execution. Besides, in the process of movement management slow movements allow so-called "feedback", the presence of which can be detected not only after the execution of a movement, but also during the execution [20,25,26]. This allows for managing movements during execution. Practising self-defence develops psychomotor abilities [27]. The basic psychomotor ability which depends on the knowledge of biomechanics is the ability to choose a proper method of defence. Coaches and PE teachers may have problems in requiring from their pupils to apply the mathematical apparatus (knowledge) in interpreting the rules of mechanics. This may be a result of the fact that the skills that pupils have acquired at school are not satisfactory and in addition they are reluctant to apply physics formulae. However, there is the possibility of using the knowledge of biomechanics in the teaching methodology of various sports disciplines. This type of the knowledge can be provided experimentally, as it was shown with aikido [2,7]. In the author's opinion this knowledge is always decently applied by PE teachers. It is advisable to incorporate a biomechanical analysis of sports techniques into teaching and not only limit to giving instructions to a pupil to lift, for example, a certain limb. Teaching an infinite number of motor sequences requires learning about the rules managing the rotary and translatory motion [2,6]. The role of a teacher or educationalist is to know these rules, because they may facilitate mastering certain motor activities by exerting an effect of understanding them based on these rules. The results obtained in experiment III show that martial arts can develop children's cognitive skills as well their motor skills [7]. In view of the above, the belief that martial arts can develop both body and mind seems justified. This, according to the author, supports Prof. E. Jaskólski's suggestion to carry out more detailed research into the way martial arts could be used in education [7, 21]. Unfortunately, there is a tendency at the moment for some of the martial arts to develop sports-wise, thus becoming ever more offensive rather than defensive [22,28]. Extreme offensiveness may border on aggression, which may result in abortion of judo philosophy and make martial arts less useful for pedagogical purposes.

5.3. The most recent research

The effect of the knowledge of mechanics on execution of aikido techniques was also analysed after testing PE students (university students, Mroczkowski A., unpublished). Similar correlations between this knowledge and aikido techniques execution as in the experiments referred to above were found. The experiment participants were also tested to find out whether understanding the rules of mechanics of a rotational movement can be related to applying aikido techniques in simplified forms of sports competition. The main objective of the experiment was to check if aikido practitioners are able to use the force of their opponents. Competition was based on similar rules to sumo, but in the apparel typical for judo competitions. Similarly to sumo, pushing a practitioner out of the bordered fight area or touching the surface with anything other than a foot body part, meant losing the game. A correlation between the knowledge of biomechanics acquired and success in the sports competition was not established (Mroczkowski A., unpublished). The results achieved could have been affected by several factors. Definitely, the short time of training, decided upon as in previous experiments [2,7], influenced the results. It is obvious that only knowledge of biomechanics is not enough for a correct execution of a certain technique with great speed and optimum dynamics. It is necessary to develop a child's correct motor features and this requires a relatively long period of time. The final success in a fight, according to Starosta and Rynkiewicz [29] is also affected by "feeling" the opponent. Thus, according to the author, the aikido forms involving a certain sports competition, sometimes similar to judo forms, should be focused on [30]. This may give a certain control over development of motor features necessary for practising in the right time techniques depending on the changing, for example, attack of the opponent. The author is presently dealing with the analysis of the effect of understanding the rules of rotational motion mechanics on mastering motor activities in rotational techniques not only in aikido, but in various sports disciplines, such as sports gymnastics, break dance and trampoline. For this purpose a training rotational simulator was used [31] (patent pending). This device is composed of two platforms on which students experience rotational movement. They can assume various body positions, for example, standing or lying. The subjects simulate the body positions assumed in sports techniques. Figure 17 shows an example of applying this device.

The whole platform on which a student is standing starts rotating. After gaining a certain velocity, the platform is no longer driven and the whole movement goes on almost without friction. A subject, for example, grabs the two bars (Fig. 17) which can imitate grasping certain parts of the opponent's body or clothes. When changing the distance between the centre of mass and the axis of rotation of the training simulator, the subject gains a change in an angular velocity according to the second law of dynamics of rotational movement. The subject is feeling the changes in angular velocity when he is changing the distance between his body and the axis of rotation. This is a good situation in which to explain to him the analogy of this movement to the aikido technique executed. Two people can practise on the rotating platform at the same time. With this device it is possible to determine the moment of inertia of the subjects if their centre of mass is located on the axis of rotation of the training simulator. For this purpose the rotating platform is driven with a falling weight. This method was already used by Griffiths et al. [32]. The initial results (Mroczkowski A., unpublished) show that the method of the experimental explanation of the principles of mechanics of the rotational movement facilitated their quicker understanding by the subjects participating in the experiment. At the same time, exercising on this device quickened the process of mastering the correct motor habits necessary for executing some aikido, break dance, trampoline and sports gymnastics techniques.

Figure 17. An example of applying the rotating training simulator

The analysis of the tests covered by the experiment raise a question: 'Isn't it worth applying biomechanics in teaching mechanics as part of the school curriculum in physics?" Biomechanics is for the author a form of so-called "live mechanics". It deals with a human being, his movements or his functioning in terms of mechanics. As the tests referred to in this paper show, this knowledge of mechanics provided in this form is well acquired by adolescents and children. It is advisable to consider whether it is possible to teach other parts of physics in this way. The author thinks that some parts of physics would prove to be useful here. For the majority of adolescents and children it would be interesting to explain the rules of their body functioning in terms of physics.

Author details

Andrzej Mroczkowski
Faculty of Physical Education at the University of Zielona Góra, Zielona Góra, Poland

Acknowledgement

I wish to thank Andrzej Kacprzak for making most of the figures included in the paper.

6. References

[1] Mroczkowski A, (2002) Integration of teaching physical education with physics. In: Koszczyc T, Oleśniewicz P, (eds.), Integration in physical education [in Polish]. AWF, Wrocław, 305–311.

[2] Mroczkowski A, (2009) The use of biomechanics in teaching aikido. Human Movement, 1(2), 10: 31–34. http://156.17.111.99/hum_mov/

[3] Bober T, (2003) Sports techniques: biomechanical approach. In: Urbaniak C. (ed.), Issues in sports biomechanics: movement techniques [in Polish]. AWF, Warszawa, 5–18.

[4] Zatsiorsky WM, (1999) Acquiring sports techniques: biomechanics and teaching. Technological progress and sports at the beginning of the 21st century [in Polish]. AWF, Wrocław, 44–46.

[5] Westbrook A, Ratti O, (1970) Aikido and the Dynamic Sphere. Charles E. Tutle, Tokyo.

[6] Mroczkowski A, (2003) Teaching aikido by using the principles of mechanics. In: Cynarski WJ, Obodyński K, (eds.), Humanistic theory of martial arts. Concepts and problems [in Polish]. UR, Rzeszów, 199–206.

[7] Mroczkowski A, (2010) The use of biomechanics in the methodology of teaching aikido to children. Archives of Budo, 6 , 2(4) 57-61. http://www.archbudo.com

[8] Walker J, (1980) In Judo and Aikido Application of the Physics of Forces Makes the Weak Equal to the Strong. Scientific American, 243(1): 126–33.

[9] Wąsik J, (2007) Power breaking in tekwondo – do physical analysis. Archives of Budo, 3: 68–71 http://www.archbudo.com

[10] Wąsik J, (2009) Chosen aspects of physics in martial arts. Archives of Budo, 5: 11–14. http://www.archbudo.com

[11] Erdmann WS, Zieniawa R, (2011) Biomechanics of judo. AWFiS Gdańsk [in Polish].

[12] Ueshiba K, (1985) Aikido. Hozansha Publishing, Tokyo.

[13] Pranin S, (1996) Interview with Koichi Tohei. Aikido Journal. http://www.aikidojournal.com

[14] Mroczkowski A, (2010) Biomechanical analysis of the aikido exercises usage for disabled people. Scientific Dissertation AWF Wrocław, 30, 83—90. http://www.awf.wroc.pl/pl/article/1003/140/Archiwum/

[15] Mroczkowski A, Jaskólski E, (2006) Effects of aikido exercises on lateral spine curvatures in children. Archives of Budo, 2, 31-34. http://www.archbudo.com

[16] Mroczkowski A, Jaskólski E, (2007) The effect of vertebral rotation forces on the development of pathological spinal curvatures. Polish Journal of Physiotherapy, 7, 1(4): 80-86. http://www.fizjoterapiapolska.pl/

[17] Mroczkowski A, Jaskólski E, (2007) The change of pelvis placement at children under influence of aikido training. Archives of Budo, 3: 1-6. http://www.archbudo.com/

[18] Mroczkowski A, (2009) The influence of the pelvis position on body posture changes. Polish Journal of Physiotherapy,. 9, 3 (4): 258—265. http://www.fizjoterapiapolska.pl/
[19] Rugloni G, (1997) Unification of Mind and Body and Ki Aikido. Ergaedizoni Genova.
[20] Bober T, Zawadzki J, (2003) Biomechanics of human motor system [in Polish]. AWF, Wrocław.
[21] Mroczkowski A, (2007) Pedagogical aspects of practicing aikido by children [in Polish]. Scientific Year's Issue Ido-Movement for Culture, 7, 103–107.
[22] Kalina RM, (2000) Theory of combat sport. COS. Warszawa, [in Polish]
[23] Kalina RM, Kalina A, (2003) Theoretical and methodological aspects of teaching lower extremity amputees safe falling. Advances in Rehabilitation, XVII: 71–79.
[24] Kalina RM, Barczyński B, Jagiełło W, et al. (2008) Teaching of safe falling as the most effective element of personal injury prevention in people regardless of gender, age and type of body build – the use of advanced information technologies to monitor the effects of education. Archives of Budo, 4: 82–90. http://www.archbudo.com
[25] Schmidt RA, (1988) Motor Control and Learning. Human Kinetics, Champaign.
[26] Schmidt RA, (1991) Motor Learning and Performance. Human Kinetics, Champaign.
[27] Harasymowicz J, Kalina RM, (2005) Training of psychomotor adaptation – a key factor in teaching self defence. Archives of Budo, 1(1): 19–26 http://www.archbudo.com
[28] Harasymowicz J, (2007) Competences of combat sports and martial arts educators in light of the holistic fair self-defence model of training. Archives of Budo, 3: 7–14 http://www.archbudo.com
[29] Starosta W, Rynkiewicz T, (2008) Structure, conditions and shaping "opponent feeling" in opinion of combat sport athletes. Archives of Budo, 4: 12–21 http://www.archbudo.com
[30] Shishida F, (2008) Counter techniques against Judo: the process of forming Aikido in 1930s. Archives of Budo, 4:4-8. http://www.archbudo.com
[31] Mroczkowski A, (2011), Rotating training simulator, Patent pending UP RP, P.395584.
[32] Griffiths IW, Watkins J and Sharpe D, (2005) Measuring the moment of inertia of the human body by a rotating platform method, American Journal of Physics, 73 (1), 85-93

Section 2

Musculoskeletal and Injury Biomechanics

Chapter 4

The Role of Skull Mechanics in Mechanism of Cerebral Circulation

Yuri Moskalenko, Gustav Weinstein, Tamara Kravchenko, Peter Halvorson, Natalia Ryabchikova and Julia Andreeva

Additional information is available at the end of the chapter

http://dx.doi.org/10.5772/50887

1. Introduction

Skull as a complicated mechanical construction, consists from 28 different bones, connected by sutures of different structural kinds - from smooth (face cranium) to teeth-like hardness (brain cranium). Basically, skull can be selected as face skull and brain skull. The last is composed from 8 bones with complicated connections between them, and namely brain skull will be described in the chapter presented.

Structure of "brain skull" (or simply – skull) is the most complicated, it is changeable and individual depending on age, sex and race. The most pronounced mechanical changes of skull properties are observed in babies and juniors which are connected with brain growing. Later, structural changes of the skull are concerned with structure of bone sutures, which tissues loss their elasticity with age. Volume of internal cavity of the skull of adult persons is balanced closely with volume of intracranial media so precisely, that the internal relief of skull bones reflects the structure of the brain and its vascular system. This confirms the idea, that skull plays protector function for brain, and it is well known.

However, this precisely high balance of internal skull volume and intracranial media volumes, due to natural variations of brain skull configuration, might be a reason of some diminishing of the skull internal volume to compare with intracranial media. Variations of the skull configuration, which actually diminish intracranial cavity, may be the most pronounced at middle ageing, when capabilities of the skull internal volume for adaptation also diminish due to decrease of the possibility of change of suture structure. Result of this may be some compressing of tissues, filled cranial cavity, first of all, liquid media blood and cerebrospinal fluid (CSF), which are responsible for circulatory-metabolic supply of brain functioning. As a consequence, some neurological symptoms may appear indicating that volume of intracranial cavity is really smaller, than optimal volume of intracranial media.

The most reasonable way to treat of such neurological disturbances is to increase of internal skull volume. That is why the first method to treat such neurological disorders was neurosurgical skull trepanation which was practiced for numerous Centuries. It is confirmed by archeological findings of human skulls with trepanation performed on living persons. The long history of skull trepanation since 2000 B.C. reviewed in many publications [1, 2, 3].

This method have been used also at middle centuries, which is confirmed by one of art master pieces, named "Treatment of foolishness" - painting of Bosch (1475), presented process of trepanation to remove "the stone of foolishness" from the skull. All these examples show, that the role of trepanation, as method of treatment of some neurological pathology, was reasoned to increase of intracranial volume, and this had been understood Centuries ago.

It is well known, that one the most important physiological function of the skull is protecting brain from external mechanical disturbances. However, some negative role of close correspondences of brain and cranial cavity volumes may be the change of balance between these volumes in different living situation, and it becomes significant, when comparative increase of intracranial volume of media is taking place. Result of this, first of all, may be decrease of activity of Cerebro-Spinal Fluid (CSF) which compresses brain blood vessels and, in some cases, brain tissue, too. All these could evoke functional insufficiencies, as headache and some neurological symptoms. The high significance of balance between intracranial cavity volume and the media filled it, is critically important in connecting with fact, that brain at all situations needs intensive and stabile blood supply, because brain metabolism is founded on aerobic principle. For aging population, the limitation of intracranial volume becomes more important and could be reason of pronounced decrease of cerebral blood flow (CBF) resulted in dementia.

Skull trepanation up to the present time is actively using when volume of intracranial media is increased by some reasons (closed brain injury, brain tumor), indicator of which is the increase of intracranial pressure.

It is necessary to mention, that so strong restrictions of intracranial cavity volume limit also possibilities for ranges of adaptation to different living situations and, diminish living capabilities. Therefore, it is reasonable to predict, that in process of evolution of mammals some protecting accommodations of intracranial volume, which are not diminish its protector function have been developed. The basis of these mechanisms should be founded on the possibility of some increase of intracranial volume compensation of change of volume media, filled the skull. Investigations at this direction become appear from the beginning of XX Century.

Primary, it was have shown, that some volume reserves is follows from the possibility of free replacements of CSF between cranial and spinal cavities and spinal cavity could accept some volume of cranial CSF due to elasticity of vertebral lumbal sack [4, 5, 6, 7]. In the middle of XX Century one more possibility, based on feature of the skull, have been discovered. It has been shown by palpation, that skull represents a complicated mechanical moveable system. This skull feature found practical application in osteopathic medicine and is used for diagnostic purposes and evaluation of results of osteopathic treatment up to the present time [8, 9]. These observations have shown, that for the skull bones the slow periodics are of special.

The next important finding have been made during observation of neurosurgical patients, which have been shown, that skull as structural unity is characterized by the special property – Cranial Compliance, which was studied by injection into the skull of neurosurgical patients of artificial (mock) CSF and dependence "Pressure/Volume" for skull has been established [10].

The next step in the evaluation of the role of skull in mechanism of cerebral circulation was made not physiologists or physicians, but mathematicians, who in process of mathematical modeling of the cerebra-circulatory system inside closed skull, have established, that only if skull will accept some additional volume of blood during systolic increase of central arterial pressure, cerebrovascular system in coupling with CSF system could function [11]. This study permits to predict, that brain circulatory support system include as active element skull bio-mechanics.

Thus, this chapter will to describe the role of peculiarities of skull biomechanics as an active component of complicate physiological mechanisms, responsible for brain physiological mechanism, responsible for circulatory-metabolic supply of brain functioning.

2. The role of skull pulse expanding in mechanism of circulatory supply of brain function

The system of cerebral metabolic supply is dependent on the interaction of a number of elements some of them are determined by the skull properties. One of the most significant in this direction is clarifying the role of CSF movement to this system inside cranium and significance at this process of the bio-mechanical properties of the skull. Critical position in the role of the skull in support of circulatory-metabolic supply of brain functioning determined by the ability of the skull to accept an additional volume of blood during the phase of systolic increase in central arterial pressure.

Indeed, arterial blood pressure consists of two components. One is the steady state of flow through the brain, determined by the basal tone of the brain blood vessels. The second occurs with each heart- beat, which initiated by arterial pressure increases and so drives the pulse volume into the cranium. It is this component that is influenced by the level of cranial compliance, which in turn depends on the volume flexibility of the skull.

This increase in arterial pressure is short – about 0.1s. This means that the cranium needs to accommodate the increase in systolic blood volume very quickly in order to use it to drive cerebral circulation. This possibility, predicted by mathematical simulation of the cerebrovascular system [12], should be follow from biomechanical properties of the skull as united bio-mechanical system.

That is important statements, because from the beginning of XIX Century up to the second part of XX Century the most of investigators belong to conception, named "Monroe-Kelly" doctrine, which on the base of majority of anatomical investigations, declare, that skull is fully rigid cavity. However, when this statement has been included to the model, it didn't function. However, when some possibility for pulse change of intracranial volume has been included to the model, it starts to work. It was the first indicator, which shows the importance of pulse change of internal skull cavity. Some years later the presence of pulse dependence on "Pressure-Volume" relation for the skull have been shown by invasive technique [13].

Follow this data, the skull could be accepted as a nearly rigid container with limited capabilities to accommodate internal volume changes in response to increases in arterial pressure. However, because investigations of Marmarou have been provided when comparative slow changes of intracranial pressure as respond to infusion to the skull artificial CSF, when volume compensation is provided also by CSF outflow to some spaces, connected with the skull by comparatively narrow gapes spinal cord, volumes, determined by arachnoid membranes of cranial nerves.

Therefore, it was important to evaluate skull itself volume reserves, in other words dynamic skull volume capabilities. Really, it is important to have the possibility to accept by the skull an extra volume of blood following the systolic increase in arterial pressure, which is in addition to the steady state level of the brain blood flow. The most rapid component of change of arterial blood pressure is evoked in systolic phase duration about 0.1s. During this time any replacements of CSF inside skull couldn't be provided [14].

There are skull expansions due to, perhaps, articular comparative mobility of skull bone in sutures of invisible value; investigations have shown, that these changes of articular bone position are less then 0.5 geometric degree [15], and this, as it follows from calculation, could increase internal volume of skull up to 3-6 ml.

Calculation of additional – systolic blood volume, which should accept cranial cavity may be provided, taking into account, than brain "portion" of stroke volume is about 10 ml. That is means, that steady stroke volume is 5 -7 ml and pulse component is 5-3 ml (Fig.1). Namely this the last volume of blood should accepted by skull during systolic phase of cardiac cycle. Comparatively to total volume of cranial cavity this volume is very small. Averaged size human brain is about 1200 ml. This value should be closely to internal volume of the skull. This means, that expand of internal volume of skull about 0.3% of initial volume to accept systolic portion of blood, could give the additional blood, which is necessary for normal functioning of circulatory –metabolic brain supply. Taking into account, that heart rate is usually 60-70 beats per minute, this additional blood volume will be about pulse evaluation due to of skull expanding could be about 110-130 ml per minute. It is known, that normal brain blood made using some relatively simple calculations. As well it was established in the middle of the last Century, brain takes 50-65ml of blood volume flows through 100g of brain mass in 1 minute [16].

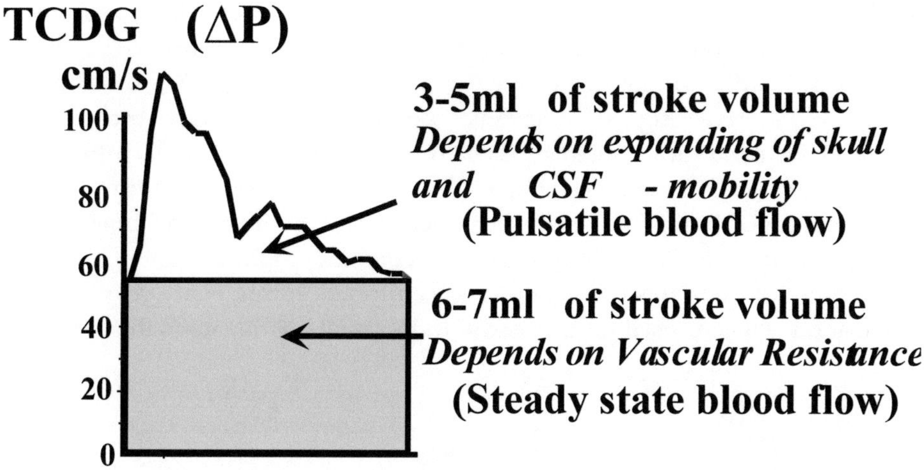

Figure 1. Distribution of steady state and fluctuating of stroke volume at normal physiological conditions.

This means, that through an average sized human brain (1200g) – about 600ml flows every minute and 110-130 ml of blood is about 20% of total brain blood supply. In case of skull is completely rigid, this additional stroke volume will by out from cerebral circulation, because additional blood couldn't inflow to the skull due to contra-pressure, determined by unchangeable volume of cranial cavity. Therefore, brain blood flow may be diminished up to 15-20%. Direct evidence of the presence of pulse skull expanding have been received by coupling of Rheoencephalogram (REG) – method, based on recording of electrical impedance between electrodes, placed in fronto-mastoid position to human head and transcranial dopplerogram (TCD) of basement of the Middle Cerebral artery [17]. REG reflect changes of intracranial blood/CSF volume, because electrical resistance of these media are significantly less to compare with brain tissue and pulse blood volume fluctuations inside skull if they are have taken place, will change common electrical resistance between electrodes [14]. TCD reflect pulse changes of linear velocity inside intracranial large arteries [18], which by Poiseuille low are proportional to volume fluctuations of these arteries, surrounded by CSF. Volume fluctuations of large brain arteries could be compensated by skull expanding due to transmitting of arterial to surrounding CSF pressure, which is a real source of forces for brain expanding.

Computed aid analysis of simultaneously recorded REG and TCD pulse give the possibility to establish "Pressure-Volume" dependence for the skull for systolic phase of pulse cycle [17], which show, that this dependence is nearly to linear and in normalized coordinates equal to line with angle to horizontal coordinate 30-40^{o}(Fig.2a) Linearity at the most cases of this dependence indicate, that pulse changes of intracranial pressure and volume are linear, which is permits to conclude, that SCF replacements inside skull during systolic part of cardiac cycle practically absent. Therefore, the fact of the presence of systolic skull pulse expansion has taken place and it is one of element of mechanism, which is responsible for circulatory-metabolic support of brain functioning and is possible quantitative to express as value of Tang. of angle of "Pressure-Volume" dependence in normalized coordinates. This dependence corresponds also to meaning "Cranial Compliance", applied to systolic phase of cardiac cycle, or in other words, "Dynamic Cranial Compliance" (DCC). Direct evidence of the role of small increasing of intracranial volume to DCC have been received by observation on neurosurgical patients just before and after trepanation, provided for the next neurosurgery [19]. These investigation have shown, that small 6-10cm^2 "window" in skull bone with saved brain cover membrane, when real intracranial volume increase to a few (3-6 ml) due to deformation of brain cover membrane, could change significantly DCC (Fig.3). The significance of this mechanism was confirmed during many years experience in skull trepanation, which is used for increase of intracranial volume up to the present time.

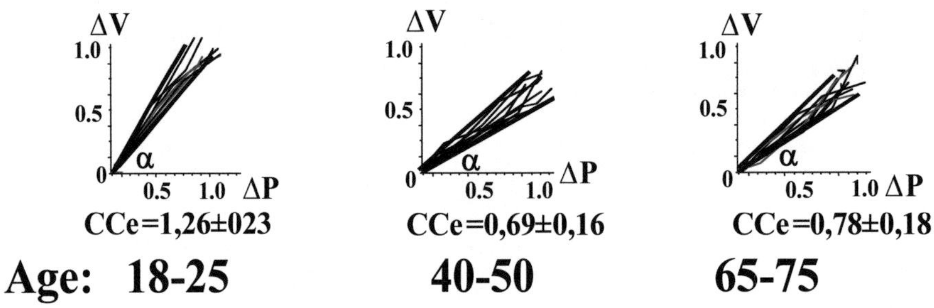

Figure 2. Averaged changes of DCC for different age groups, showing the decline in CCe in the middle-age group.

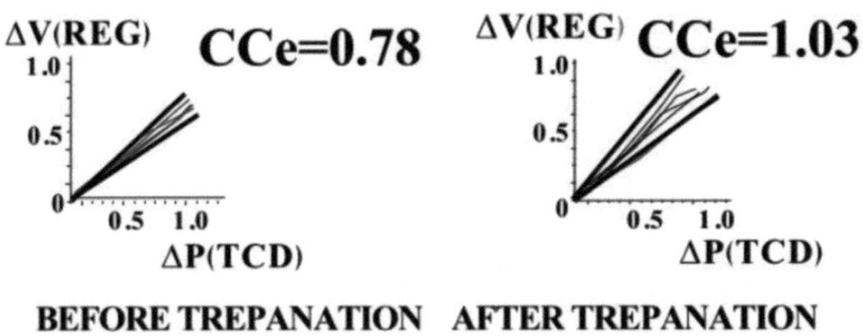

Figure 3. Averaged DCC values before and after trepanation. Changes in the biomechanical elasticity of the skull allow_the cranium to accept additional blood with each pulse stroke.

All above data permits to conclude, that without property of rapid expanding of the skull when DCC =0, brain blood flow will decrease with the same indices of arterial and venous central pressure to significant value – up to 15-20%., which is significant for brain functioning. The known data shows, that even some less significant decrease of DCC may in some cases reflect to brain activity. This suggestion is confirmed by observations, which shows, that DCC gradually decrease up to age 40-50 and, then increase again-Fig. 2b,c [20]. That abnormal decrease of angle of normalized curve widely vary for different persons and its pronounced values are correlate with some neurological symptoms (headache, decreased working capabilities). It is important to mention, that an increase of DCC after 55-60 Years (Fig.2c) is not connected with the skull properties. It is determining by aging decrease of brain mass, which have been shown by MRI investigations [22]. Investigations, provided with aging persons, after 65-70 by comparing results of blood flow measurements, and DCC measurements and determination level of dementia level by psychophysiological computerized method "Prognosis", show (Fig. 4), that level of decrease of cognitive brain function and DCC are proportional but not closely correlate with values of level of brain blood that even small [17]. It is possible to conclude, that biomechanical properties of the skull determine its expanding due to pulse increase of intracranial pressure permits to accept additional volume of blood during systolic increase of central arterial pressure, with play sometimes significant role in supporting circulatory-metabolic supply of brain functioning.

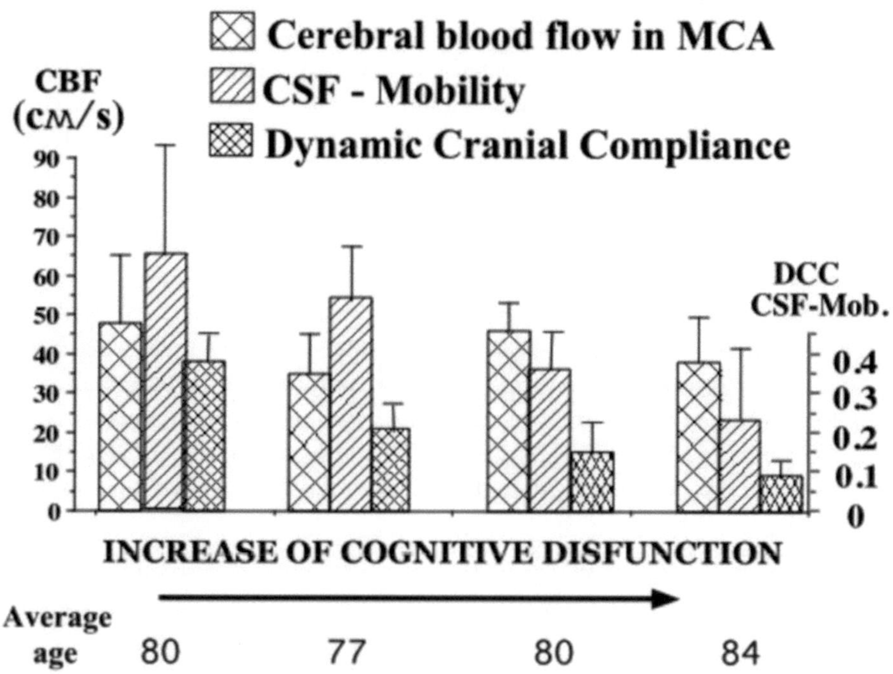

Figure 4. Dependence between value of DCC , Cerebral blood flow and CSF – mobility in aging human groups which are different by level of cognitive disfunction.

During diastolic part of cardiac cycle biomechanical properties of the skull also play role in supporting of cerebral blood circulation due to energy, collected during systolic increase of central arterial pressure. In this phase the role of CSF replacements is increase. In the beginning, they, in coupling with skull biomechanics, provide the distribution of pulse blood volume inside skull and, then, support pulse outflow of venous blood from the skull.

Thus, evaluation of DCC for the skull during each cardiac cycle is comprised of the initial interval as a rapid and nearly linear increase of arterial pulse pressure which lasts from 0.05 - 0.15s and perfectly reflects "Pressure-Volume." dependence, or DCC, which is determined by the equivalent elasticity of the cranium due to the biomechanical of the skull structure elements

It is important to emphasize that changes of the steady component of brain blood flow, determined by the perfusion pressure, are independent of the pulsatile component. The total brain blood supply is determined by the superposition of the steady state perfusion pressure (average level of arterial pressure) and the components of blood flow, which are in turn determined by the biomechanical properties of the skull and the mobility of cerebrospinal fluid within the skull.

3. Slow fluctuation of skull bone motion

Data, which could be predict, that there are slow fluctuations of skull bone have been known more than a Century. However, the first direct observation concerned skull bone motion has been received by palpation of human skull [23], and it developed the special branch of medicine, - "Osteopathy", founded by Dr. A. Still, and actively spreading in USA since the end of XIX Century. This phenomenon consists in periodical with changeable amplitude and frequency skull bone movements, in range 6-12 cycles per minute. The high similarity of these fluctuations with respiratory movements was a reason to name these fluctuations as "Primary Respiratory Mechanism", or PRM. Decades of studies of PRM by palpation permits to describe peculiarities and consequences of involving of particular skull bone to motion, which are regular at normal conditions and different for pathology. During this time a number of conceptions and hypotheses of physiological nature of PRM have been formulated to explain the origin of these phenomena [24, 9]. Some of them were unreal from positions of biophysics and physiology, but at the present time there are acceptable conceptions which could be regarded as working hypothesis. From classical osteopathic positions, initial point of PRM is liquids disturbances in cranium, which slightly move the brain and acts to brain membrane in region of occipital and basis bone. This initiates movements of other skull bones.

One of the explanations of peculiarities PRM was based on reciprocal tension of brain membranes, which is popular up to the present time. However, brain membranes have no contractive elements and this is a current problem to accept of this conception, but skull membranes could play the role of passive "modulator" [25], which determine connection between movements of particular skull bones. Summary, the acceptable the conception, founded on the fact, that for cerebrovascular system is typical periodical changes of vascular tone. The consequent of these changes is intracranial pressure fluctuations, which may be a real physical force for deformation of skull pattern as united biomechanical system. Combination of fluctuations of intracranial pressure with additional role of skull membranes as passive modulator looks at the present time the most acceptable conception for slow periodical skull bone motions.

The experimental study of the skull bone motions started with cadaver observation, where skull bone motions were initiated by saline injection into the skull. These investigations give negative results. Later it was understood, that postmortem changes in sutures make skull as a solid body. Then instrumental investigations of living skull bone motion in animal experiments have been fulfilled [26, 27]. Human observation under physiological conditions on the base of modern technology, have been appeared at the end of XX Century. Firstly, direct observations demonstrated, that skull represents a complicated mechanical moveable system. This suggestion is based on investigations, represented device with needles inserted through cover skull tissues in human head and fixed in skull bone. Invisible movements of upper end of the needle were registered by means of small mirror fixed there, which are deflected laser beam, focused to the mirror [28].

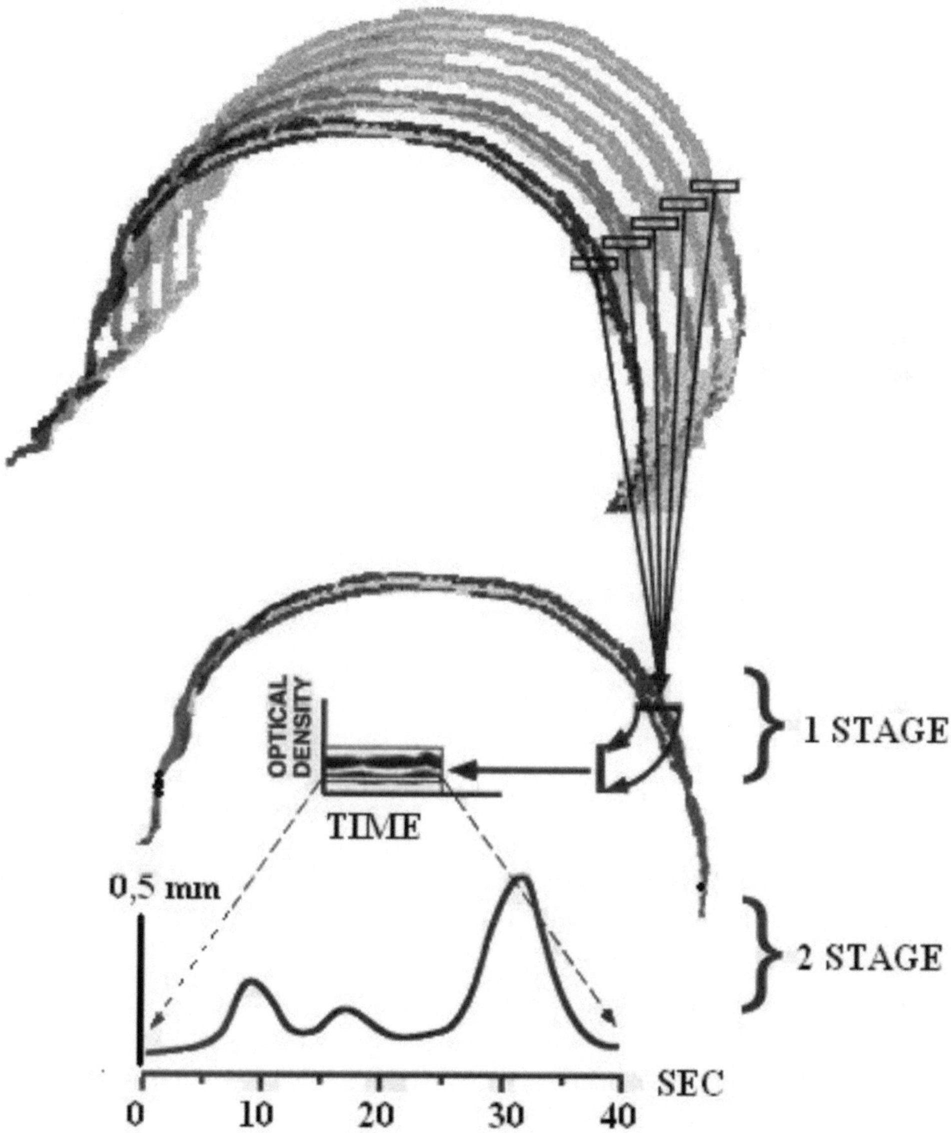

Figure 5. Principle of transforming individual regions of image series into the amplitude-time plot: (a) input of the image series into a computer and specification of the analyzed region; (b) collocation of images and creation of the intermediate image; and (c) transformation of the intermediate image into the amplitude-time plot.

Approximately at the same time investigations, based on skull bone image analysis have been provided, using method of image computer analysis of serial MRI or X-Ray pictures – 30-45 single shots. Serial images of the skull were recorded by means of the Siemens-PolyStar angiographic system in 23 patients Principle of this method have been based on inserting to computer memory a number of equal fragments of skull bone image and, after an increase of their contrast superposition, small deviations of position of these fragments in united skull it is possibly as time dependent graph.(Fig. 5) These passive observation have shown, that, for normal physiological conditions, movements of skull fragment images, selected at of both MRI and X-Ray pictures, are periodical with irregular amplitude, fluctuating in ranges about 0.1-0.4 mm and frequency 5-12 cycles per minute (Fig.3 and Fig.6a).

However, skull bone motion should be recorded it in conditions, closely to real physiological experiment, when skull bone motions are evoked by some external procedure, which may be fixed by intensity of intervention and its duration. Such conditions could be provided by observation of skull bone movements during angiographic procedure, taking for analysis cases with absence of clear pathology in cerebrovascular tree. During angiographic procedure, when into the skull through Internal Carotid Artery 20 ml of X-Ray contrast solution during one sec. is injected with pressure significantly higher than arterial pressure. During such procedure, about 15-18 X-Ray shots were made, which permits to evaluate by image analysis of skull bone motions with amplitude up to 0.7-0.9 mm, at the end of the phase of increasing of intracranial volume evoked by injection of solution, and decrease after 2.5-3.0 sec, when X-Ray solution has passed through brain vascular tree, and intracranial volume normalized (Fig.6). Taking into account, that average volume of intracranial cavity is about 1200ml, its increase on peak of X-Ray solution injection will be 1.0-1.5%. Taking this value into account it is possibly to suggest, that slow skull bone articular periodic fluctuations, accepted firstly manually and later instrumentally, represent about 0.2 – 0.4% of intracranial volume. At the present time stages of these skull motions are described in details [29, 30]. The fact, that skull bone movements are reciprocal, have been confirmed by simultaneous recording of REG with "lobe-occipital"(REG1) and "bi-temporal"(REG2) position of electrodes. How it follows from Fig 6 Graph, with REG1-REG2 coordinates, that received two-dimensional pictures rather wide, which may be if comparative distance between electrodes is changed. That means, that every couple of electrodes moves reciprocally (Fig.7)

Thus, at the present time is confirmed, than slow skull bone motion are taking place and their motions are reciprocal. Similar slow fluctuations involve spinal cavity, too. In PRM phase of increasing volume of skull initiate to replacement of some volume of CSF to spinal cavity, which is possible, because volume-pressure changes are slow enough and CSF returned to the skull, when intracranial pressure decrease. The fact which confirmed this statement, have been received in experiments with animals [31] and demonstrated reciprocal slow volume changes in the skull and spinal cavity. Similar observations in humans have been received recently [21].

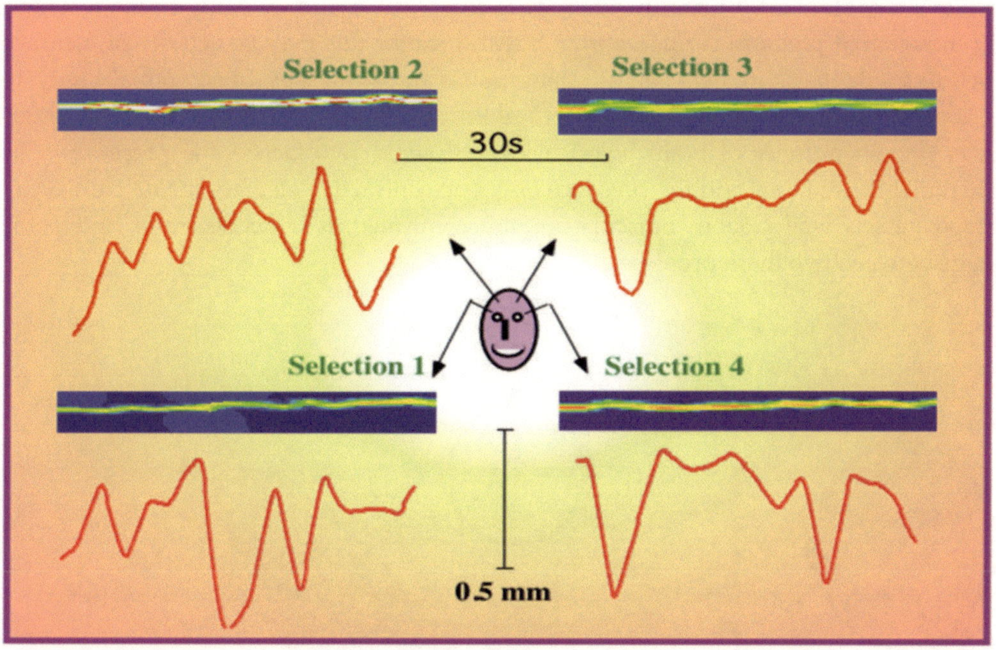

(a)

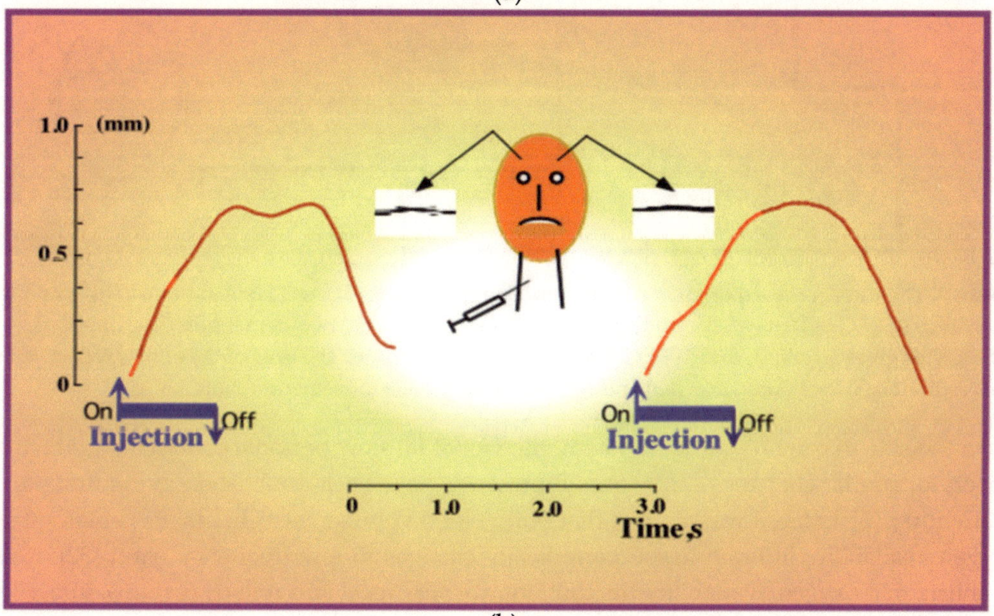

(b)

Figure 6. Fig.6 Cranial bone movements obtained by serial X-ray imaging (a) in a physiological state and (b) in the case of injection of radiopaque solution. The analyzed sections are shown in the center. (A) Consecutive image series showing changes in section shadows in the specified regions of cranial bones. (B) Changes in cranial bone positions in the specified sections shown in the amplitude-time plot. In all curves, the starting points of the time count are brought into coincidence and the time scale is the same.

Mechanism of slow skull bone fluctuations is complicated and its study needs to establish why intracranial pressure is fluctuating. A real force for this may be activity of contractive structure inside the skull. Between different tissues and structures, filled cranial cavity, only one is capable of active change its mechanical properties due to external source of energy – that is smooth muscles of brain blood vessels. Because two facts – the presence of slow intracranial fluctuations and the presence only one contractive element inside cranial cavity – blood vessels wall smooth musculature, are confirmed, it is necessary to find possible linkage between two these processes.

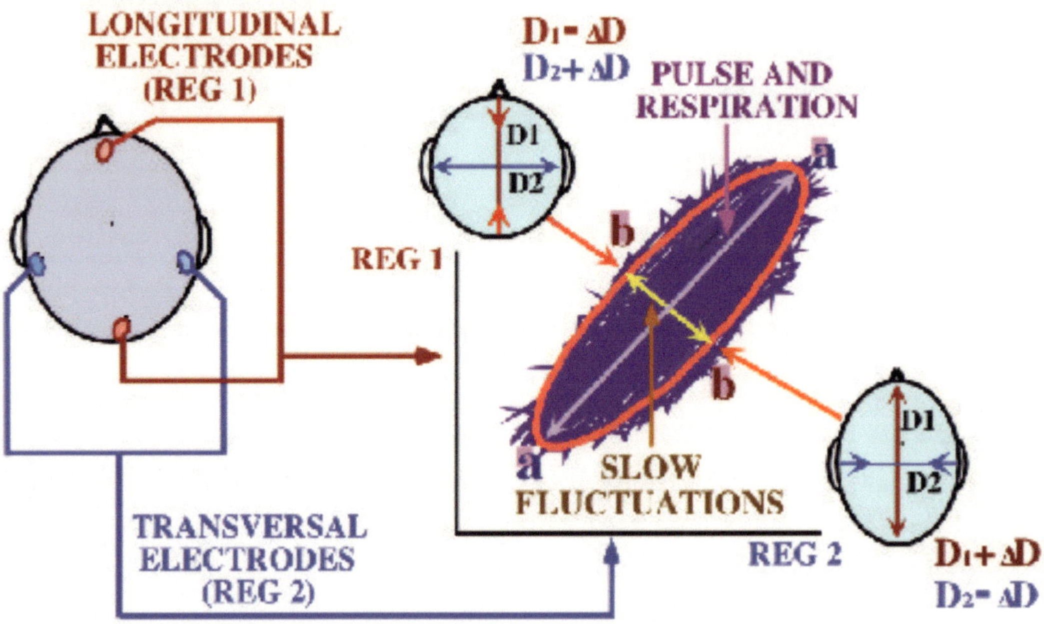

Figure 7. Diagram of the experiment which demonstrates reciprocal skull bone motions. The 2-min recordings of REG by "cross" electrode position are represented on two-demention diagram (dark violet). This gives an ellipse due to REG fluctuations superposition. The axes of the ellipse reflect pulse and respiratory waves (long axe a-a) and reciprocal skull bone movements (short axe b-b).

From the side of cerebrovascular system, the fact of its slow periodic contraction, which are reason for similar changes of intracranial pressure, have been established in the first part of XX Century [7]. Later, it was shown, that brain blood volume, recorded by REG method and oxygen availability in brain tissue periodically changes in low frequency band [32]. It was establish, that oxygen availability fluctuations are very local and reflect, perhaps, in nervous tissue metabolic processes, but REG fluctuations reflect comparative wide brain region and show changes of brain blood volume. With purpose to find the correlation between this and other slow fluctuations, which are special for intracranial media, simultaneous recordings of REG in both hemispheres, TCD and chest movements were provided at a group of healthy persons 20-30 Years. It has been shown, that spectrum of all these processes is characterized by three kind of fluctuations (Fig 8). The first, the most pronounced peak is heart pulsation.

REG changes during this peak are one more confirmation of the skull pulse expanding. The second peak is corresponds to respiratory movement of chest. One more peak – low frequency corresponds to similar fluctuations, recorded by TCD and corresponds to slow fluctuation of central arterial pressure – Traube-Hering-Mayer waves. Between this peak and peaks, reflect chest respiratory movements, only on REG spectrum it is possible to see some one – three peaks, which didn't correspond to any peaks of respiratory and TCD records and belong to REG only. This is permits to think, that these peaks represent the origin of slow intracranial volume fluctuations, which corresponds to slow cranial bone motions, named as PRM. Generally, mechanism of slow fluctuation it is possible to present as scheme shown on Fig.9, which demonstrate, that very small, below one geometric degree comparative fluctuations of skull bone position could significantly change internal volume of cranial cavity and source of forces for these changes are fluctuations of intracranial pressure of different origin, mainly due to vascular tone fluctuations.

The next and perhaps the last question in analyzed chain is the origin of slow cerebrovascular fluctuations, depended mainly on vascular tone. It is not yet definite answer to this question. However, it looks real prediction, that the fluctuations reflect control processes in the cerebrovascular system, because to vascular wall continuously acts different factors, everyone of which could change vascular tone. This is, first of all, different kinds of innervation – adrenergic, cholinergic, peptidergic and purinoergic nature [33], different mediators, nonorganic ions, autocoids, change of intravascular pressure and others [34]. Simultaneous, under normal living conditions, acting to vascular wall of numerous factors is the most real reason of appearing of some non-regular fluctuating process, which could be reason of intracranial pressure fluctuating, evoked periodical skull bone movements.

However, it is not only one reason for slow changes of intracranial pressure fluctuations, which may have connection with skull bone fluctuation. One of cranial osteopathy positions describes the special phenomenon, called crania-sacral rhythm. This is means the reciprocal movements of sacral section of vertebral column and the skull. Explanation of this phenomenon, given by osteopathy is not acceptable from point of view of biomechanics. Recently a new explanation, based on two facts has been appeared. One of these facts follows from MRI observation of pulse CSF movements in sagittal section of skull and neck. The second is the data of REG, taken from sacral region of vertebral column.

If compare images, which have taken every 0.1s, it is possible to see that some portion of pulse CSF volume don't return to the skull and moves along to vertebral cord to its lumbar sack. REG records show, that cranial and sacral pulsations have reverse phases and level of REG gradually change. These data permit to formulate hypothesis, that during every pulse cycle some amount of CSF fills lumbar sack and pressure increased. Because hydrostatic forces are strong, this increase of lumbar sack could slightly erect sacral region of vertebral column, which is possible to feel by palpation. The erection may stimulate around sacral section muscles, which return it to initial position and CSF is returning back to cranium. This is, may be not perfect but some, basing on observation explanation of crania-sacral rhythms, observation of which is "classical" method of osteopathy [35].

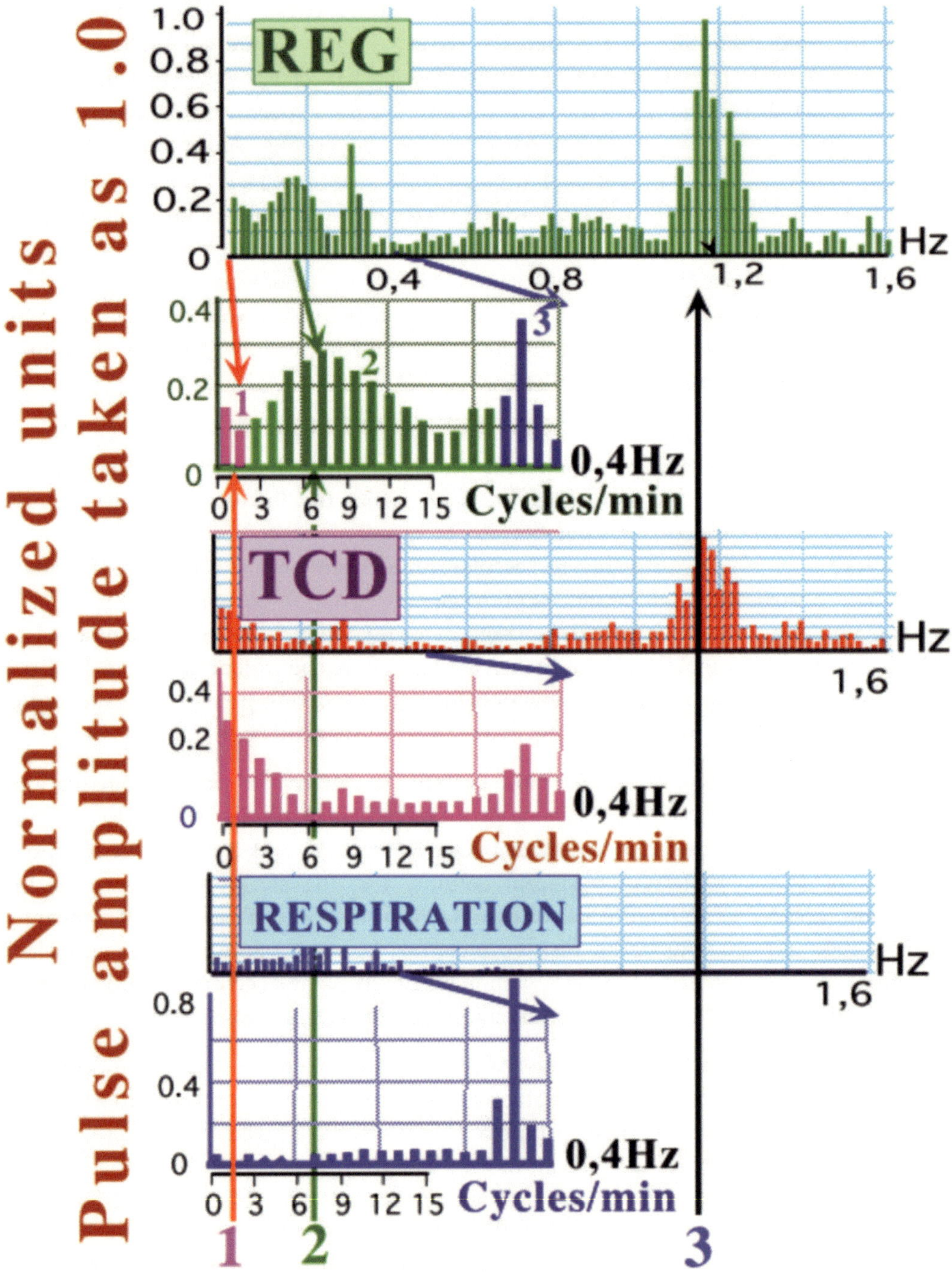

Figure 8. Spectral representation of healthy person, age 23, slow in range 0-1,6 Hz and range 0.04 Hz of REG, TCD and Respiration. Arrow (RED) show off central arterial pressure, Arrow Green – fluctuations of intracranial origin and BUE arrow show component of chest respiratory movements.

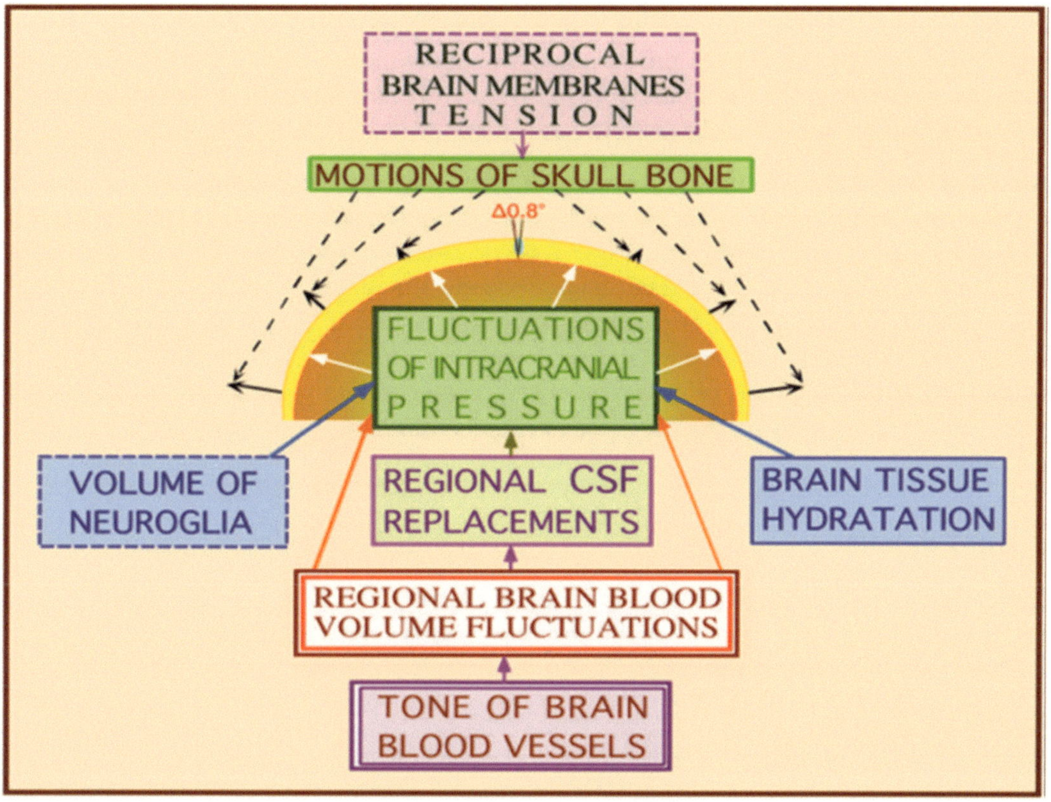

Figure 9. Schematic representation of skull move motion, iniciated intracranial pressure , roun indicate physiological origin (BROW) but may be pathological origin – (BLUE).

4. Conclusion

All above presented data show, that skull, its brain part is a moveable system. For the skull bone two kinds of movement are special. The first is rapid, during 0.1-0.15s skull expanding by pulse increased arterial pressure. This is too short time to involve CSF to balance intracranial blood volume and increase of volume of arteries mainly on skull basement increase pressure into the skull, which causes deforming its pattern by articular movements of bones in sutures which totally increase internal volume of skull by not much value – about 0.2-0.3% to compare with total volume cranial cavity. However, it is enough to accept additional 3-6 ml of blood, which are extremely important for maintain circulatory-metabolic supply of brain functioning. If this kind of skull mobility diminishes by some reason, brain circulatory insufficiently could appear, which may be reason of some such pathology situation, as brain dementia of circulatory origin. That is why on early steps of civilization, skull trepanation served as method of treatment.

Other kind of the skull bone movements is based on slow fluctuation of skull bone position, evoked by internal forces, finally expressed as periodical changes of intracranial pressure, which include processes in vertebral cord also. Actually, cranio-spinal space is represented as nearly united system. Intracranial liquid system is an initial movement of this slow fluctuation process. It is not yet final point of view how developing all connected with this processes and origin of this initial liquids movement is not clear yet. However, the fact of slow skull bone motions is looks definite confirmed.

Although the current role in skull bone movements play mechanical properties of sutures, because separate bone of skull are mechanically too strong to be deformed by arterial pressure forces or other origin forces, which occur inside cranial cavity. Thus, final result of the skull expanding of slow skull bone motions are depends on not structure and biomechanics of some particular suture, but of skull, as united complicated bio-mechanical system. Biomechanical properties single elements composed skull as united moveable system and this is a new property, which is appeared on systemic level. Mechanical properties of the united mechanical system may be different, to compare with any single elements.

Author details

Yuri Moskalenko*, Gustav Weinstein and Julia Andreeva
Institute of Evolutionary Physiology and Biochemistry, Russian Academy of Sciences, Sankt Petersburg, Russian Federation

Tamara Kravchenko
Russian School of Osteopathic Medicine, Moscow, Russian Federation

Natalia Ryabchikova
Biological Faculty Moscow State University, Moscow, Russian Federation

Peter Halvorson
ITAG, PA, USA

5. References

[1] Mednikova MB. 2004. Skull trepanation in ancient times. Aleteya: Moskow. 206p.
[2] Mogle P, Zias J. 1995. Trephination as a possible treatment for scurvy in a middle bronze age (ca. 2200 B.C.) skeleton. Intern. J. of Osteoarchaeology. V.5 p.77-81.
[3] Jenkner F. 1966.Prähistorische und präcolumbianische Schädeltrepanationen. Klagenfurt: Kartner Druckerei. 36p.
[4] Cushing H, Studies in Intracranial Physiology and Surgery, London, 1926.
[5] Sepp EK, Die Dynamik der Blutzirkulation im Gehirn, Springer, Berlin, 1928.
[6] Moskalenko YuE, Naumenko AI, 1957. About oscillatory movements of CSF in craniospinal cavity. Physiol. J. USSR. V.43. No.10. p.928-933.

* Corresponding Author

[7] Moskalenko YuE, 1967. Dynamics of the brain blood volume under normal conditions and gravitational stresses. Nauka Press, Leningrad. (English translation: NASA-TT F-492).

[8] Chaitov L. 1999. Cranial manipulation. Theory and practice. Oseus and soft tissues approaches. Churchill-Livingstone, London. 293p.

[9] Kravchenko TI, Kusnezova MI. 2004. Cranial osteopathy. St.Peterburg. 78p. (In Russian)

[10] Marmarou A, Shulman K, LaMorgese J. 1975. Compartmental analysis of compliance and outflow resistance of the cerebrospinal fluid system. J Neurosurg v.43. p 523–534.

[11] Menshutkin VV. 2010. The skill of biological modeling (ecology, physiology, evolution). Karelian Branch of Russian Acad. Sci.. Petrozavodsk-St.Petersburg. 419p.

[12] Moskalenko YuE, Kisliakov YuYa, Weinstein GB, Zelikson BB. 1972. Biophysical aspects of the intracranial circulation. Amer. Heart J. v.83. No.3. p.401-414.

[13] Marmarou A, Shulman K, Rosende R, 1978. Nonlinear analysis of CSF system and intracranial pressure dynamics, J Neurosurg v.48 p.332–344.

[14] Moskalenko YuE, Weinstein GB, Demchenko IT, Krivchenko AI, 1980. Biophysical Aspects of Cerebral Circulation, Pergamon Press: Oxford.

[15] Moskalenko Yu, Frymann V, Kravchenko T, Weinstein G. 2003. Physiological background of the cranial rhythmic impulse and the Ptimary Respiratory Mechanism. The AAO Journal. V.13. No.2. p.21-33.

[16] Kety SS, Schmidt CF, 1945. Cerebral blood flow in man. Amer. J. Physiol. v27. p.53-66.

[17] Moskalenko Y.E., Ryabchikova N.A., Weinstein G.B. et al 2011.Changes of circulatory-metabolic indices and skull biomechanics with brain activity during aging. J. of Integrative Neuroscience. V.10. No.2. p. 131-160.

[18] Aaslid R, 1986. Transcranial Doppler sonography. Springer-Verlag. N.Y.

[19] Moskalenko Yu E, Weinstein GB, Kravchenko TI, Mozhaev SV, Semernya VN, Feilding A, Halvorson P, Medvedev SV, The effect of craniotomy on the intracranial hemodynamics and cerebrospinal fluid dynamics in humans, *Hum Physiol* 34:299–305, 2008.

[20] Moskalenko YuE, Weinstein GB, Halvorson P, Kravchenko TI, Ryabchikova NA, Feilding A, Semernia VN, Pqanov AA 2008. Biomechanical properties of human cranium: Age-related aspects. J. of Evolutionary Biochemistry and Physiology. V.44. No.5. p.513-520.

[21] Moskalenko YuE, Weinstein GB, Halvorson P, Kravchenko TI, Feilding A, Ryabchikova NA, Semernia VN, Panov AA, 2007. Age peculiarities of relationships between brain blood flow, liquor dynamics and biomechanical properties of human cranium. Russian Physiol. J. V.93. No.7. p.788-798.

[22] Courchesne E, Chisum HJ, Townsend J, Cowless A, Covington J, Egaas B, Harwood M, Hinds S, Press GA, 2000. Normal brain development and aging: Quantitative analysis of in vivo MR imaging in healthy volunteers. Radiology. V.216. No.3. p.672-682.

[23] Sutherland WG, 1939. The Cranial Bowl. A Treatise Relating to Cranial Mobility, Cranial Articular Lesions and Cranial Techniques,. Free Press Co, Mankato, MN.

[24] Chaitov L. 1999. Cranial manipulation. Theory and practice. Oseus and soft tissues approaches. Churchill-Livingstone, London. 293p.

[25] Moskalenko Yu.E., Kravchenko T.I. 2004. Wave phenomena in movements of intracranial liquid media and the Primary Respiratory Mechanism. The AAO Journal . v.14. No.2 p.29-40.

[26] Fryman V. 1971. A study of the rhythmic motions of the living cranium. J. Amer., Osteopathic Ass. V.70. No.5. p.928-945.

[27] Adams T., Heisel RS, Smith VC, Briner J. 1992. Parietal bone mobility in the anesthetized cat. J Amer. Osteopathic Ass. V.92. No.5. p.599-611.

[28] Livandovski MA, Drasby E, Morgan V, Zanakis MF 1996. Kinetic system demonstrates cranial bone movements about the cranial sutures. J. Amer., Osteopathic Ass. V.96. No.9. p.552.

[29] Moskalenko Yu, Frymann V, T. Kravchenko T, G. Weinstein G. Physiological mechanisms of slow fluctuations inside cranium. Osteo (La revue des osteopathes. France). Part I. 1999. No.50. p.4-15. Part.II. 2000. No.51. p.4-11.

[30] Moskalenko Y, Weinstein G, Kravchenko T, Gaidar B, Semernia V. 1999. Periodic mobilitynof cranial bones in humans. Human Physiology. V.25. No.1. p.51-58.

[31] Moskalenko YuE, Naumenko AI, 1959. Investigation of CFS translocations in normal animals. Physiol. J. USSR. V.45. No.5. p.562-568.

[32] Moskalenko Yu, Cooper R, Crow H, Walter WG, 1964. Variation in blood volume and oxygen availability in the human brain. Nature. v.172. No.4928. p.159-161.

[33] Moskalenko YuE, Beketov AI, Orlov RS. 1988. Regulation of the cerebral circulation: Physical and chemical ways of investigation. Nauka: Leningrad. 160p. (In Russian)

[34] Demchenko IT, 1983. Blood supply of the awake brain. Nauka Press, Leningrad. 180p (in Russian).

[35] Moskalenko YuE, Kravchenko TI, Weinstein GB, Halvorson P, Feilding A, Mandara A, Panov AA, Semernia VN. 2009. Slow-wave oscillation in the cranio-sacral space: A hemoliquorodynamic concept of origination. Neuroscience and Behavioral Physiology. V.39. No.4. p.377-381.

Chapter 5

Biomechanics of the Fractured Femoral Neck – The New BDSF-Method of Positioning the Implant as a Simple Beam with an Overhanging End

Orlin Filipov

Additional information is available at the end of the chapter

http://dx.doi.org/10.5772/47839

1. Introduction

The femoral neck fracture is subjected to powerful shearing forces due to the angular, spiral-like architecture of the proximal femur. Under the conditions of severe *osteoporosis*, the *femoral neck* consists of cortical walls, enveloping soft cancellous bone, having unimportant mechanical significance, and the neck can often be looked at as a hollow cylinder. If the condition of patient is not appropriate for total hip replacement (mental diseases or other risks), and a decision is made for a screw fixation, the implanted screws must be solidly fixed in the distal fragment in at least two points in order to provide resistance to the shearing forces in case of osteoporosis. The traditional screw fixation methods, however, do not meet the above-named requirement. Present-day popular *traditional methods* of femoral neck fixation, which are performed by three cancellous screws, placed parallel to each other and parallel to the femoral neck axis, are associated with poor results in 20 to 42% [1,2,3,4,5]. The high failure rate of traditional screw fixation methods can be explained by the presence of a number of related biomechanical imperfections. (1) *Instability of the construction regarding varus stress.* The entry points of the three screws in traditional screw fixation methods are located at the thin, fragile cortex of the greater trochanter or close to it. The screws are often placed in the soft cancellous bone near the axis of the femoral neck, with no cortical support [6]. Even if one or two of the distal screws are placed close to the distal cortex of the femoral neck, they lack any second solid point of support. A second point of support for them is the thin and fragile lateral cortex of the greater trochanter – their entry point. Such a construction can rely only on the interfragmental compression, generated by the intraoperative tightening of the screws, but the achieving of compression depends on the solidity of the cancellous bone. This circumstance results in high failure rate in cases of osteoporosis. (2) *Lack of sliding phenomenon*. Upon body weight loading, in the process of the subsequent wedging of the osteoporotic cancellous bone in the fracture site, the screws of traditional methods of fixation, which lack two-point cortical support, cannot effectively slide distally and laterally keeping unchanged their angle towards the axis of diaphysis, and rather have expressed tendency to displace in varus, with fixation failure. (3) *Inability to move the entry point of the screws distally into the solid diaphyseal cortex, and simultaneous placing of three parallel screws*. In 1961 Garden [7], like other authors before, further developed the concept that the implants must have more vertical placement, similar to the direction of the medial compression lamellae of the femoral neck internal trabecular system, in order to provide resistance to the shearing forces. However, when developing this concept, the classic authors used only one implant (a nail). Thus the implant successfully provided resistance to the shearing forces, but it did not create compression between the fragments, because it is not a screw and it is also not able to ensure reliable rotational stability of the head fragment, being only one [8,9,10].

The anatomy of proximal femur does not allow simultaneous placing of three screws, which are parallel to each other, and lie near the cortex in the periphery of the femoral neck and, at the same time, have their entry points positioned distally, in the solid cortex of the diaphysis, in order to avoid the fragile lateral metaphyseal cortex.

When applying the *conventional* methods of positioning three parallel screws, in case movement of the screw entry point distally is attempted, the screws will be placed at a very obtuse angle towards the diaphysis and obliquely to the femoral neck. By increasing the angle of penetration, the surface of the femoral neck cross-section decreases geometrically, and in practice the placement of more than one or two screws is hard to be accomplished. However, a two-screw fixation does not provide reliable stability in all planes [10]. Both problems are resolved by the newly introduced method of *Biplane Double-supported Screw Fixation* trough the concept of biplane positioning of the implants.

The *Biplane Double-supported Screw Fixation Method in femoral neck fractures (BDSF-method)*, developed by O. Filipov, is a new method of screw fixation in femoral neck fractures, based to *an original concept of the establishment of two supporting points for the implants and their biplane positioning* in the femoral neck and head. The console-like proximal femur requires the fixation screws to have to support the weight-bearing head fragment, acting like a *beam with an overhanging end,* which must have two points of support in the distal fragment. The concept of the Biplane Double-supported Screw Fixation (BDSF) method is based on this principle. This method is original with the three screws being laid in two planes, which aspect allows the entry points of two of the implants to be placed much more distally, in the solid cortex of the proximal diaphysis, and also to lean onto the strong femoral neck distal cortex. Thus, we establish two points of support. The achieved by this method position of the distal screw and the middle screw as well, in view of statics, turns them into *a simple beam with an overhanging end, loaded by a vertical force*. This beam with an overhanging end successfully supports the head fragment, bearing the body weight and transferring it to the diaphysis.

2. Biplane double-supported screw fixation method - Operative technique

Indications: Fractures of the Garden types from I to IV. The indications and contraindications for application of the BDSF-method are generally the same as of the conventional methods for fixation of the femoral neck fractures. The BDSF-method significantly expands the indications for application of screw fixation in terms of the bone, changed by osteoporosis. The method is most useful and has no reasonable alternative in adult patients above 80 years with a high cardiopulmonary risk, as in patients with severe concomitant diseases, inclusively some mental diseases, patients with senile dementia, and others, for which the primary joint replacement may be contraindicated.

Implants: 7.3 mm self-tapping cannulated screws

Reduction: Mild traction, slight abduction and internal rotation of the limb are applied. Only anatomical reduction is acceptable.

Approach. A straight lateral incision, starting at the level of the lower border of the greater trochanter, with distal length of 6 to 10 cm. A stripping of the periosteum of the lateral diaphysis at 6-7 cm is performed.

Placement of the implants. When applying the BDSF-method, the three cannulated screws are placed in the frontal plane at a highly increased angle. Both the distal and the middle screws touch tangentially on the curve of the distal femoral neck cortex (Fig.1-a.). At internal rotation of the leg, in anteroposterior view, the projection of the distal screw usually crosses the projections of the other two screws, thus forming the letter F (F-technique). Via the concept of *biplane positioning*, developed by the BDSF-method, the three screws are placed in two vertical oblique planes (in lateral view). The two planes diverge towards each other in the direction of the femoral head, and are oblique towards the frontal plane. The distal screw is laid in the dorsal oblique plane. The middle and the proximal screws are placed in the ventral oblique plane (Fig.1-b.).

First of all, we lay the guiding wire for the distal cannulated screw. Its tip is placed at 5-7 cm distally from the lower border of the greater trochanter in the anterior one-third of the surface of the stripped off diaphysis. It is directed proximally at an angle of 150 – 165° towards the diaphyseal axis, with inclination from anterodistally to posteroproximally, so that after it touches tangentially on the curve of the distal femoral neck cortex, the wire goes into the dorsal third of the femoral head.

The middle guiding wire is placed secondly. The entry point is at 2 to 4 cm proximally from the entry point of the distal wire, but in the dorsal one-third of the stripped off surface of the diaphysis. This wire is placed at an angle of 135-140° towards the diaphyseal axis and inclined from posterodistally to anteroproximally, so that after it touches tangentially on the curve of the distal femoral neck cortex, the wire goes into the front one-third of the femoral head. In the frontal plane (anteroposterior view) the tip of this guiding wire goes into the distal one-third of the femoral head.

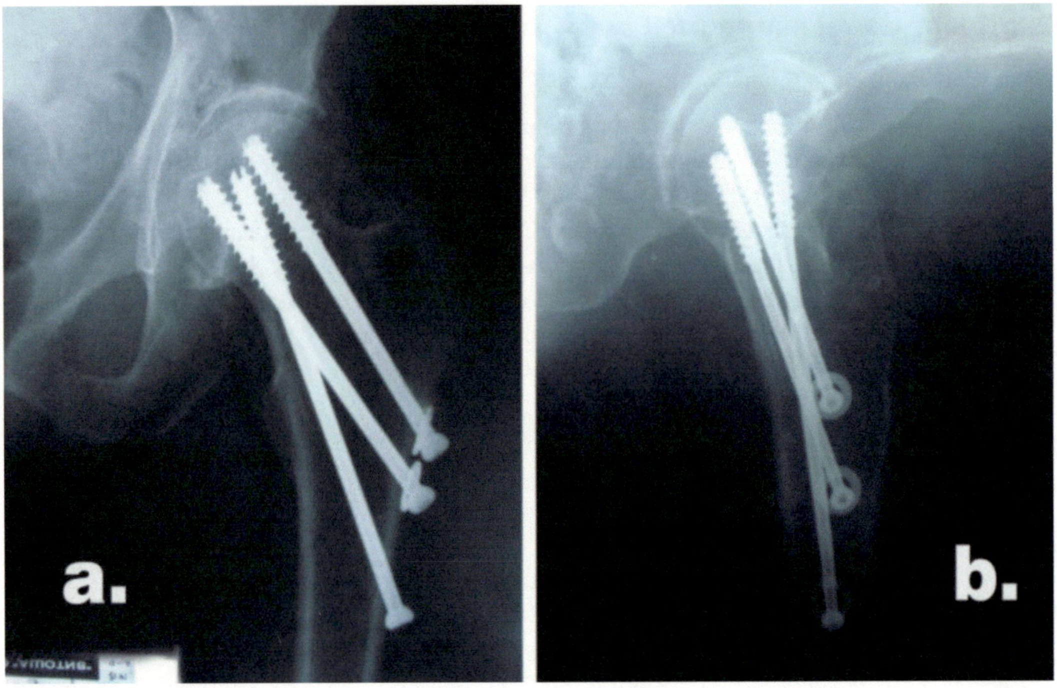

Figure 1. Radiography. a. Anteroposterior view; b. Lateral view.

Last to be laid is the proximal guiding wire, with its entry point at 1-2 cm proximally from the entry point of the middle wire, in the dorsal one-third of the stripped off diaphysis, close to the lower border of the greater trochanter. Placed parallel to the middle wire, the proximal wire also goes into the front one-third and into the proximal one-third of the femoral head.

Afterwards we drill and place the screws one by one. Before placing the middle and distal screws, we overdrill their holes in the thick lateral cortex by using a 7.0 mm cannulated reamer.

The middle and the proximal screws are placed first, because they are perpendicular to the fracture surface. Next we release the foot traction, and a repeated impaction of the fracture with an additional tightening up of the screws follows. We perform the impaction gently by hammering on a plastic impactor on the diaphyseal cortex. Finally, the distal screw is placed.

The guiding wire easily changes its initial direction when passing through the thick diaphyseal cortex, therefore its tip is guided into the necessary direction by the surgeon's free hand with the help of a cannulated instrument.

When we place the two distal guiding wires tangentially over calcar femorale, if we suddenly feel resistance, it indicates that the tip has got into the medial diaphyseal cortex distally from the femoral neck. In such a case we change the direction by increasing the angle of penetration. For this purpose we use the same hole in the lateral cortex, by taking out the wire completely and after a change in the direction of the wire, we try by the high-speed rotating trocar tip to change the direction of the channel in the cortical hole, in order to achieve the necessary angle of the wire through the bone. Sometimes, it is necessary to leave the old hole and to bore anew next to it. Sometimes, although very rarely, that is not sufficient and release from the thick lateral cortex is required, by reaming around the placed in poor position guiding wire by the 4.5 mm cannulated reamer. Thus the wire is freed from its contact with the lateral cortex in the created opening of 4.5 mm and is easily directed in the necessary direction.

The placing of screws under very oblique angle requires following of the principle for their two-plane positioning and none of the screws must be placed in the central part of the femoral neck and head (in profile projection). If some of the screws are placed in the central part of femoral neck, it will be an obstacle for placing of the other two screws. The distal screw, which is more obliquely placed, must be located in the posterior part of the femoral neck and head. If in violation of BDSF-technique it is placed in the anterior part of the femoral head, then because of the physiological anteversion of the femoral neck, it will be difficult or impossible for the other two screws to be placed in the posterior part of the femoral neck, because there is a tendency to find them too marginally at the cortex of the femoral head.

Radiographic time: from 0.2 to 0.3 minutes

Mean operative time: 39 min (30 to 45 min)

Postoperative period. Limited weight bearing for 4-6 months, by using two crutches would be perfect, if the patient is cooperative.

3. Biomechanical basis of the BDSF-method

This method's innovation is laying of the three screws in two planes, which allows for the entry points of two of the implants to be placed much more distally, in the solid cortex of the proximal diaphysis, and also to lean onto the femoral neck distal cortex. Thus we establish two points of support. The solid distal cortex of the femoral neck acts as a medial supporting point for the screws, which works under pressure - *supporting point A*. The entry points of two of the screws (the distal and the middle one) in the thick cortex of the diaphysis, ensure a second solid supporting point for the screws – a lateral one, which works under tension (or pressure in proximal direction) - *supporting point B*. The position of the distal screw as well as of the middle screw, which are achieved by the method, in terms of the *statics*, turns them into *a simple beam with an overhanging end, loaded by a vertical force.* This beam with an overhanging end successfully supports the head fragment, bearing the body weight and transferring it to the diaphysis. Furthermore, due to the biplane placement, enough space for a third screw is provided, unlike the classic authors' models, where just one or maximum two implants are placed at an obtuse angle [7,11]. Another advantage of the method is that due to the increase in the distance between the two supporting points, the weight borne by the bone is significantly reduced *(see the static analysis)*. An advantage of the BDSF-method is that the entry points of the screws are positioned wide apart from each other, which ensures that upon weight bearing, the tensile forces spread over a greater surface of the lateral cortex and thus the risk of its fracturing decreases significantly. Another advantage with the BDSF is that the screw, placed at a highly increased angle, works in a direction close to the direction of the loading force, which guarantees better results for the screw in its role as a beam because the influence of its sagging decreases.

4. Static analysis

With the *conventional methods* of femoral neck fixation - by three cancellous screws, placed parallel to each other and parallel to the femoral neck axis, the entry points of the three screws are placed at the thin, fragile cortex of the greater trochanter or close to it. The screws are often located near the axis of the femoral neck in the soft cancellous bone, without any cortical support. According to my previous investigations [12], with *conventional methods*, due to the lack of two solid supporting points, the implant acts statically like a *beam on an elastic foundation.* The elastic foundation is implemented by the soft cancellous bone (Fig.2).

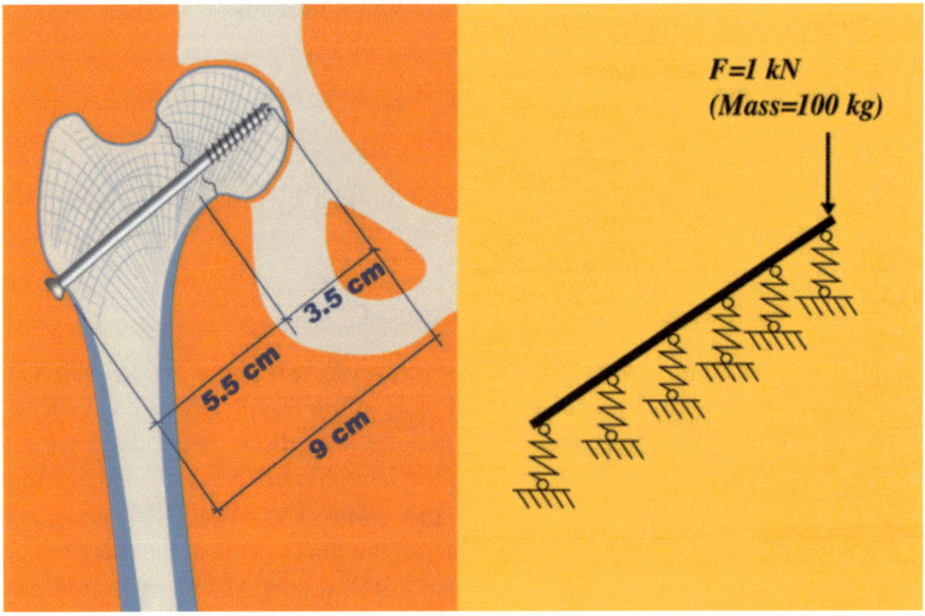

Figure 2. Static model of the conventional methods of fixation – the implant acts statically like a beam on an elastic foundation. *F = load;*

In contrast to the conventional methods, when the *Biplane double-supported screw fixation - method* is applied, the implant is additionally supported at points **A** and **B** of the cortex. The interaction between the implant and the cancellous bone is neglected, because of the comparatively small stiffness of the cancellous bone. In this way, with enough practical accuracy, with the BDSF-method, the static model is considered to be *a simple beam with an overhanging end* (Fig.3). This beam is supported at points **A** and **B** only.

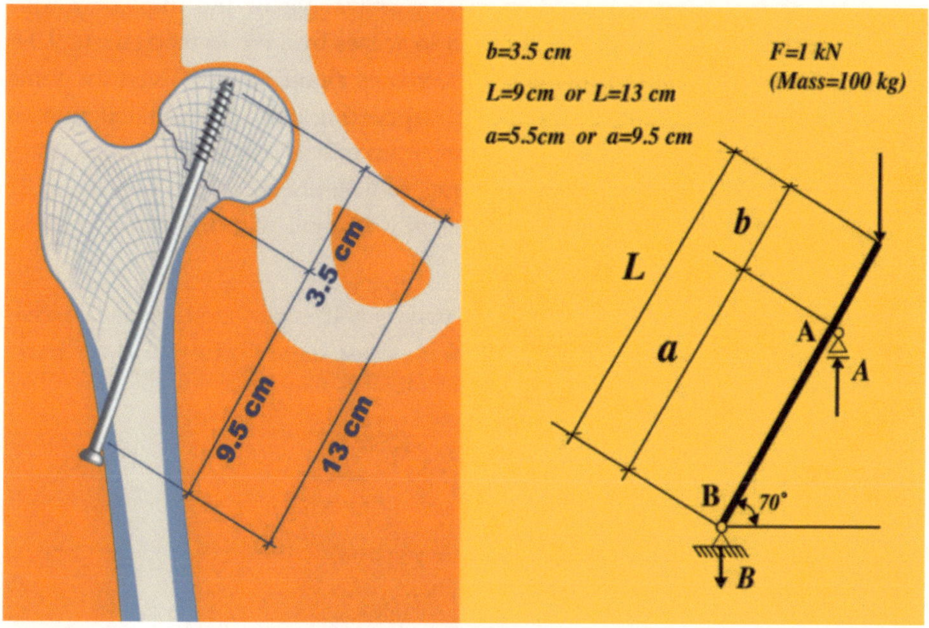

Figure 3. Static model of the new BDSF-method of fixation – the implant acts like a simple beam with an overhanging end. *F = load; L = length of beam; a = distance between points A and B*

Applying the well-known equilibrium equations for a beam, we obtain the forces acting on the cortex at supporting points **A** and **B**.

The load acting at point **A** is pressure in a distal direction and it equals to $A = \frac{FL}{a}$;

The load acting at point **B** is pressure in a proximal direction and it equals to $B = A - F$.

At the BDSF-method, due to the increase in the distance between the two supporting points, the weight borne by the bone is reduced. If we look at two cases of equal vertical weight but different distances between the supporting points, we will see that the greater the distance, the smaller the weight at each of the two supporting points.

The average anatomical distance from the tip of the screw to the distal femoral neck cortex curve (**point A**) is 3.5 cm (Fig. 4.).

With *conventional methods* **(case 1.)** the average distance from **point A** to the entry point of the screws in the lateral cortex (**point B**) is 5.5 cm (a = 5.5 cm). In order to make a comparison with the BDSF, when body weight of 100 kg is given, with conventional methods the load acting on the curve of the femoral neck distal cortex (if the screws lean on this support at all) is estimated as

A equal to 1.63 kN (163.63 kg). The load on the fragile lateral cortex (**point B**) is estimated as **B** equal to 0.63 kN (63.63kg), directed in the opposite direction (proximally).

With *the BDSF method* **(case 2.)**, with increasing the angle of the implant towards the diaphysis, the distance between points **A** and **B** increases by 4 cm to reach up to 9.5 cm (a = 9.5 cm). That is why, the load on the cortex decreases significantly. Given the same body weight of 100 kg, the load acting on the medial supporting point is estimated as **A** equal to 1.36 kN (136.84 kg) or with *16.38% less than conventional methods*, and on the lateral supporting point the load is estimated as **B** equal to 0.36 kN (36.84 kg) or with *42.11% less than conventional methods*. The distal screw normally applied with the BDSF method has a length of 13 cm.

The lateral cortex stress state around point **B** is complex. It is subjected to compressive stress in a proximal direction, and to horizontal tensile stress as well. In the lower part of the cortex the stress is mainly tensile.

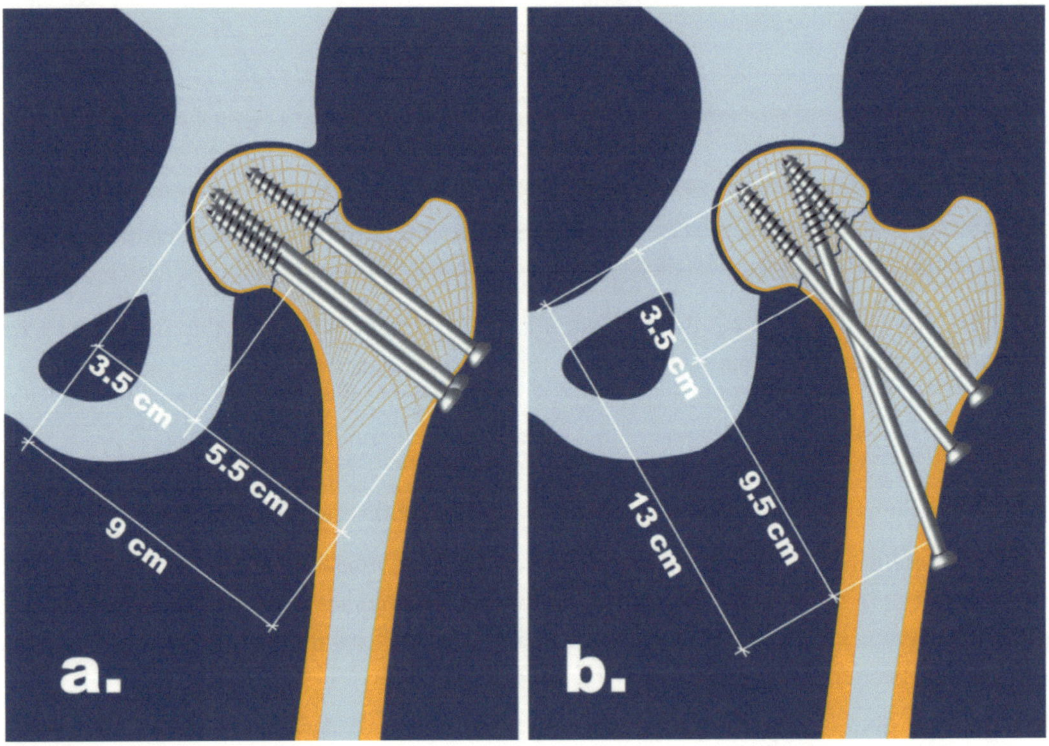

Figure 4. Fixation of the femoral neck: a. Conventional method; b. The BDSF-method. [12]

These forces of tension are responsible for the occurrence of subtrochanteric fracture as a complication of the screw fixation. As it was mentioned, these forces of tension are decreased by 42% with the BDSF-method, compared to the conventional methods of fixation. Besides, with the BDSF-method the entry points of the screws are located wide apart from each other (from 2 to 4 cm), which leads to dispersion of the tension stress on the lateral cortex over a wide surface and decrease of the fracture risk, contrary to the conventional methods, with which the entry points of the screws are at a distance less than 1 cm from each other and the forces of tension are concentrated over a small surface.

5. Clinical results of the BDSF-method

The BDSF-method was introduced in 2007 and it was applied by different surgeons since than. From a series of 178 patients, who underwent surgical treatment, 88 were studied [12]. Inclusion criteria was having late control x-rays and examinations after discharge with questionnaires filled-in. Out of the 88 studied patients, 27(30.68%) were male and 61(69.31%) - female patients; the average age was 76.9 (with the youngest patient at the age of 38 and the eldest at the age of 99). Grouping patients by age: 18 patients (20.45%) were at the age of under 69; 27 patients (30.68%) were at the age of 70 to 79; 37 patients (42.04%) were at the age of 80 to 89; 5 patients (5.68%) were at the age of 90 to 95 ; 1 patient (1.13%) was aged 95 to 100. More than one concomitant disease, which influences the results of Harris Hip Score, was found in 21 patients (23.86%). The average follow-up period is 8.06 months.

The Garden classification was used for classifying of the fractures as follows:

Garden type I: 3 (3.41%); Garden type II: 1 (1.14%); Garden type III: 9 (10.23%); Garden type IV: 75 (85.02%).

Results. From the studied 88 patients fracture union was registered in 87 patients (98.86%) and failure in 1 patient (1.13%).

Assessment according to the Harris Hip Score (modified): Poor results – in 10 patients (11.36%). Fair results – in 20 patients (22.72%). Good results – in 21 patients (23.86%). Excellent results – in 37 patients (42.04%).

The average Harris Hip Score is 84.26 points [13].

6. Unusual and difficult cases

Difficult for management are the unstable fractures and the fractures with vertical fracture line Pauwels type III.

Unstable fractures. In the elderly patients, at the age above 80, the preoperative reduction is usually achieved easily because of the fact that the fracture occurs upon low-energy trauma and although it seems displaced at a diagnostic X-rays (Garden III and IV), the fracture is usually stable and in the process of reduction there is a good control over the head fragment.

In younger and active patients the fracture usually occurs with more severe traumatic influence, for example falling over slippery surface, falling from a greater height (from stairway or in road accidents). In these cases more severely expressed tearing of soft tissues around the fracture occurs frequently and the fracture is severely displaced. Following the incident some patients try to get up and step on the limb, thus causing additional displacement of the fracture or additional fragmentation, which turns one banal fracture of the femoral neck into a an unstable fracture. When there is a severe displacement of the fracture, clinically the patients are with more expressed external rotation and shortening of the limb and have a history of more severe traumatic influence, or patients report for attempts of getting up and stepping, followed by repeated falling. At a diagnostic X-ray the usually registered grade according to Garden is type IV+ with severe external rotation of the distal fragment. In these cases frequently is found that the distal fragment "hangs" at the fracture table on the lateral view under its own limb weight. In such cases the reduction and fixation can turn to be extremely difficult and a doubtful prognosis of the femoral head survival can be assessed. It is reasonable in such patients if they are not at a young age a decision to be made for a primary joint replacement. In the presence of contraindications for joint replacement, if, nevertheless, a decision is made for metal fixation, we try the usual preoperative reduction: traction, abduction and internal rotation or sometimes a reposition by Leadbetter. If the preoperative reduction is not successful, we use frequently the intraoperative reduction, as under the conditions of sterility, the hanging distal fragment is lifted by the surgical assistant or by a special attachment of the fracture table. With achieving of reduction we use the guiding wires for temporary fixation of the fracture, followed by screw fixation. The intraoperative reduction is a procedure with a high risk for failure and the beginning of the surgery without a successful preoperative reduction of the femoral neck fracture frequently is followed by an open reduction.

There exists a group of *unstable* fractures, with which the proximal fragment is too rotated and stands in valgus position, with fracture surface directed laterally. A frequent cause for this is the V-shaped fracture surface with presence of a spicule, which is obstructive to the reduction. If after an attempt for reduction on the fracture table by abduction, traction and internal rotation, the fracture reduction remains unsatisfactory, we apply a developed by the author method for reduction by traction, abduction, external rotation, release of traction, internal rotation and adduction.

Filipov's technique: The traction is increased, the limb is abducted and externally rotated in order to wedge away the fragments of the vicious position; next a complete release of the traction is applied and thus the distal fragment skips the obstacle and comes into contact with the head fragment placed in valgus in a new mode. Then internal rotation is applied and adduction of the limb, with the distal fragment reaching the head fragment in anatomical position or achieving reduction.

Sometimes the unstable fractures of the femoral head require open reduction.

The fractures with vertical fracture line (*Pauwels type III*) are difficult for metal fixation. If the patient is with contraindications for primary joint replacement (young age) and it however requires metal fixation, the popular method of choice are the implants with fixed angle [14]. In these fractures the curve of the femoral neck distal cortex is included to the proximal fragment and it makes inefficient the fixation with screws alone. In vertical fracture line a good fixation is achieved with the 130º blade-plate, placed low in the distal one-third of the femoral neck. In order to be avoided fracture displacement during the placing of the blade of the plate, I recommend, following placing of the guiding wire for the blade of the plate, to be performed preliminary fixation of the fracture with one cannulated 7.3 mm screw, placed in the upper one-third of the femoral neck, parallel to the guiding wire for the blade of the plate. The fixed angle of the blade of the plate successfully counteracts to the shearing forces and its double-L cross-section counteracts to the torsion forces until reaching of healing. An alternative technique is a valgus accomplishing osteotomy at the level of lesser trochanter, with fracture surface placed into more horizontal plane and shearing forces turned into compressive. The fixation is with a DHS-plate or with 130º blade-plate. For the Pauwels type III fractures in the present are used successfully *locking plates.*

7. Other popular methods for fixation of the femoral neck

The present methods for fixation of the femoral neck are two types: fixation with cancellous screws and fixation with massive implants with fixed angle.

Methods for fixation of the femoral neck with cancellous screws. At present different methods for screw fixation are used, with typical for the conventional methods placement of the screws parallel to each other and parallel to the axis of the femoral neck. The most popular are the methods with three parallel screws, placed in a configuration of a triangle – two screws distally and one screw proximally; the inverted triangle configuration; the configuration of four parallel screws, placed with square-like form; configuration of three parallel screws, situated in one plane vertically. The main goal in all of these methods is achievement of compression between the fragments. Besides there is a striving of placing the screws with divergence in the femoral head. It is recommended the screws to be placed as far as possible in the periphery, close to the cortex, in order to be achieved maximum stability of fixation. Fixation with screws is also popular, connected with a small side plate.

The presented new method of *Biplane Double-Supported Screw Fixation* of the femoral neck provides new opportunities in biomechanical and clinical regard, which surpasses all known up to now methods of screw fixation for this fracture.

Alternative implant systems. Massive implants with a fixed angle.

DHS (AMBI)-plate. The fixation with DHS (AMBI)-plate is considered as an alternative method for screw fixation in fractures of the femoral neck. An advantage of the DHS-plate is the fixed angle, which ensures support of the femoral head in regard to the varus stress. Disadvantages of the femoral neck with DHS (AMBI-plate), especially in the presence of osteoporosis are as follow: (1.) The DHS-screw fixes the fracture only in one point and usually requires placing of one additional, antirotational screw, which severely increases the volume of metal, implanted in the femoral neck; (2.) Upon loading, the 135-degree DHS implant not always ensures effective sliding-phenomenon, and with severe osteoporosis the body weight loading sometimes leads to cutting of the DHS-screw through the soft cancellous bone of the femoral head with migration of the implant in proximal direction, accompanied by displacement in varus of the fracture; (3.) Increase of the percentage of aseptic necrosis with fixation with DHS, compared to the screw fixation. (4.) In the presence of osteoporosis, the compression upon the fracture, created by the DHS-screw is very weak, compared to the three cancellous screws, which fix in the head subchondrally in three different points.

Proximal femoral locking plates. These implants represent a modification of the traditional methods for fixation with cancellous screws, placed almost parallel in the cancellous bone of the femoral neck. Here the screws are fixed in the plate at the level of the lateral cortex, which solves the problem with the fragile lateral cortex of the greater trochanter and creates a stable construction. However it fixes the fracture statically, not allowing creation of compression, because of the locking of the screws and also lacks the sliding-phenomenon, which is helpful for the process of healing.

Intramedullary nails. In the presence of femoral neck fracture, combined with other fracture, located in a lower segment of femur, at present we use different types of intramedullary systems of the type of the reconstructive nail and PFN.

Other alternative types of implants, most of which have only historical significance, are the 130° blade-plates. Their inconvenience is that they cannot create compression as the screw systems and having at the same time imperfections with their outdated surgical technique. However, having a fixed angle the blade-plate ensure excellent fixation of the fracture regarding the varus stress and torsion and combined with one additional screw is probably the most effective method for fixation in fracture with vertical fracture line – type Pauwels III.

8. Conclusion

The provision of two steady supporting points for the implants and the obtuse angle at which they are positioned, allows transferring of the body weight successfully from the head fragment onto the diaphysis, owing to the strength of the screws, with the patient's bone quality being of least significance. The position of the screws allows them to slide under stress at minimum risk of displacement. The achieved results with the BDSF-method in terms of fracture consolidation are far more successful than the results with conventional fixation methods. The BDSF-method ensures reliable fixation, early rehabilitation and excellent long-term outcomes, even in non-cooperative patients. BDSF is mainly addressed to patients, who have contraindications for arthroplasty, as well as for conventional screw fixation.

Author details

Orlin Filipov
Department of Orthopaedics, Vitosha Hospital, Sofia, Bulgaria

9. References

[1] S.E. Asnis, L. Wanek-Sgaglione (1994) Intracapsular fractures of the femoral neck. Results of cannulated screw fixation. J Bone Joint Surg. 76, Vol.12, pp. 1793-1803

[2] R. Blomfeldt, H. Törnkvist, S. Ponzer, A. Söderqvist, J. Tidermark (2005) Internal fixation versus hemiarthroplasty for displaced fractures of the femoral neck in elderly patients with severe cognitive impairment. J Bone Joint Surg. Br 87-B, Vol.4, pp. 523-529

[3] J.E. Gjertsen , T. Vinje, L.B. Engesaeter, S.A. Lie, L.I. Havelin, O. Furnes, J.M. Fevang (2010) Internal screw fixation compared with bipolar hemiarthroplasty for treatment of displaced femoral neck fractures in elderly patients. J Bone Joint Surg. Am 92, pp. 619-628

[4] G.L. Lu-Yao, R.B. Keller, B. Littenberg, J.E. Wennberg (1994) Outcomes after displaced fractures of the femoral neck. A meta-analysis of one hundred and six published reports. J Bone Joint Surg. Am 76, Vol.1, pp.15-25

[5] J. Tidermark, S. Ponzer, O. Svensson, A. Söderqvist, H. Törnkvist (2003) Internal fixation compared with total hip replacement for displaced femoral neck fractures in the elderly. J Bone Joint Surg. Br 85-B, Vol.3, pp. 380-388

[6] S. Lindequist (1993) Cortical screw support in femoral neck fractures. A radiographic analysis of 87 fractures with a new mensuration teqhnique. Acta Orthop.64, Vol.3, pp. 289-293

[7] R.S. Garden (1961) Low-angle fixation in fractures of the femoral neck. J Bone Joint Surg. Br 43-B, Vol.4, pp. 647-663

[8] L. Hernefalk, K. Messner (1996) Rigid osteosynthesis decreases the late complication rate after femoral neck fracture. Archives of Orthopaedic and Trauma Surgery, 115, pp. 71-74

[9] V. Selvan, M. Oakley, A. Rangan, M. A-Lami (2004) Optimum configuration of cannulated hip screws for the fixation of intracapsular hip fractures: a biomechanical study. Injury 35, Vol.2, pp. 136-141

[10] E. Walker, D. Mukherjee, A. Ogden, K. Sadasivan, J. Albright (2007) A biomechanical study of simulated femoral neck fracture fixation by cannulated screws: effects of placement angle and number of screws. Am J Orthop. 36, Vol.12, pp. 680-681

[11] J. Dickson (1953) The "unsolved" fracture: a protest against defeatism. J Bone Joint Surg. Am 35, pp. 805-822

[12] O. Filipov (2011) Biplane double-supported screw fixation (F-technique): a method of screw fixation at osteoporotic fractures of the femoral neck. Eur J Orthop Surg Traumatol 21, pp. 539-543

[13] W.H. Harris (1969) Traumatic arthritis of the hip after dislocation and acetabular fractures: treatment by mold arthroplasty. An end-result study using a new method of result evaluation. J Bone Joint Surg. 51A, pp. 735-755

[14] F. Liporace, R. Gaines, C. Collinge, G. Haidukewych (2008) Results of internal fixation of Pauwels type-3 vertical femoral neck fractures. J Bone Joint Surg. Am 90, pp. 1654-9

[15] Tidermark J, Ponzer S, Svensson O, Söderqvist A, Törnkvist H (2003) Internal fixation compared with total hip replacement for displaced femoral neck fractures in the elderly. J Bone Joint Surg. Br 85-B(3): 380-388

Chapter 6

Comparison of Intracranial Pressure by Lateral and Frontal Impacts – Validation of Computational Model

Aalap Patel and Tarun Goswami

Additional information is available at the end of the chapter

http://dx.doi.org/10.5772/50368

1. Introduction

Traumatic brain injury (TBI) is a leading cause of death in the United States. The brain is among the most essential organs of the human body. From a mechanical stand point, different scenarios where a head comes in contact with a media has evolved a number of integrated protection devices. The scalp and skull, but also to a certain extent the pressurized subarchnoidal space and the dura matter, are the natural protections for the brain. However, these structures are not adapted to the dynamic loading conditions involved in modern road and sports accidents as well as blast injury scenarios. The consequence of this extreme loading is often moderate-to-severe TBI [1-15]. Injuries to the head constitute one of the major causes of death. Brain injury disables or kills someone in the United States every two and half minutes [2]. The annual hospitalization and rehabilitation cost has been estimated to be $33 billion per year in the US alone [14]. In the United States TBI is a leading cause of death for persons under age 45 [15]. TBI occurs every 15 seconds, see Figure 1. Approximately 5 million Americans currently suffer some form of TBI disability. The leading causes of TBI are motor vehicle accidents, falls, sports injuries and from blast injuries [12]. Thus, preventing these head injuries will not only enhance safety and quality of life but also save healthcare dollars.

Over the last 40 years, biomechanical research has been gaining attention to fully understand the mechanism of the head injury. Understanding and thus protecting the brain from injury. This can only be achieved by: 1) understanding mechanics of the impact and 2) the biomechanical response of the head to a variety of the loading conditions [2]. A cost-effective alternative method using the finite element modeling was used to investigate TBI of human head subject to impact loadings [3].

A brief review of TBI performed below and injury parameters compiled for model validation.

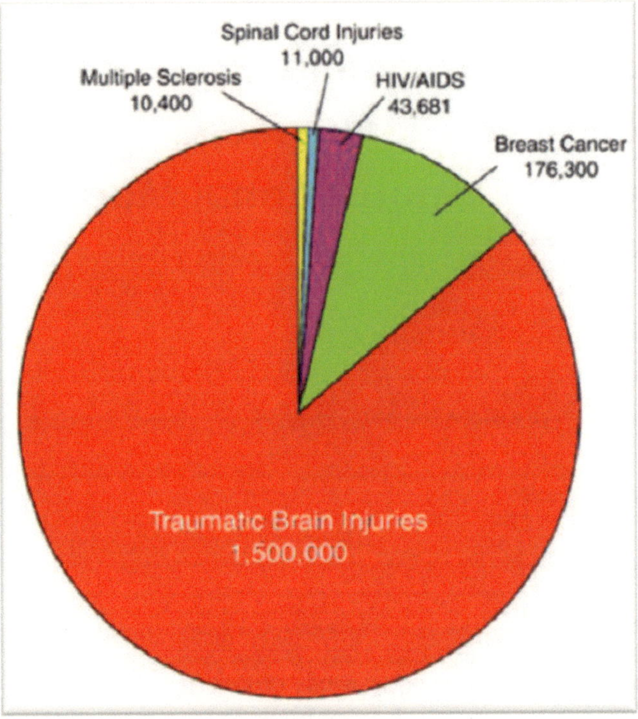

Figure 1. Scope of traumatic brain injury in comparison to other modalities [13]

2. Head injuries

The human head consists of three components [9],

1. The bony skull - Cranial and facial bones
2. The skin and other soft tissue covering the skull. Which consists of layers known as the SCALP (**S**kin, **C**onnective Tissue, **A**poneurosis (Galea), **L**oose connective tissue and **P**eriosteum
3. The contents of the skull. Most notably the brain, but also including the brain's protective membranes (meninges) and numerous blood vessels, shown in Figure 2

Injuries to the skin may be categorized as superficial or deep, and include contusion (bruise), laceration (cut), and abrasion (scrape). Injuries to the skull may break one or more of the bones of the skull in which case the skull is said to have been fractured (broken). Two aspects of a skull fracture are 1) whether it is open, or 2) depressed [10]. Injuries to the brain and associated soft tissue are the result of either head impact or abrupt head movement (e.g., deceleration injury) or some combination of the two. Injuries may be due to the skull fracturing and being pushed inward (a depressed fracture), or from the brain impacting the interior of the skull, or from internal stressing of the brain (i.e., shear, tension and/or compression). The complexities of the head and brain systems are reflected in head injury consequences, Figure 3.

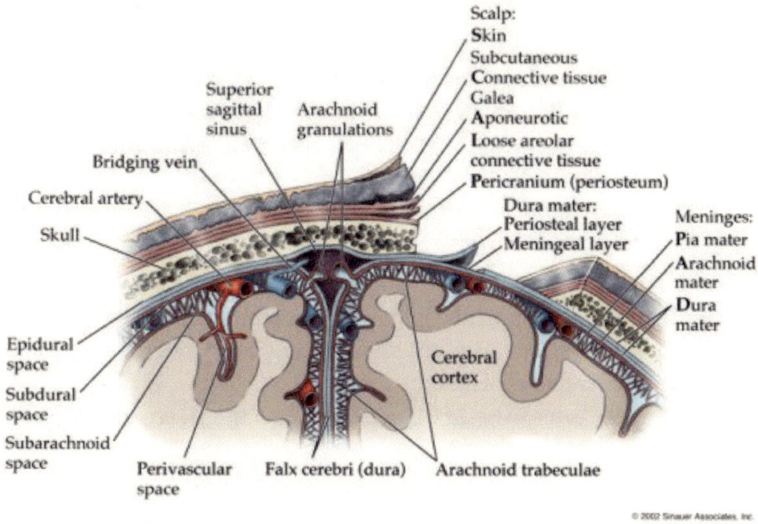

Figure 2. Anatomy of the human head [10]

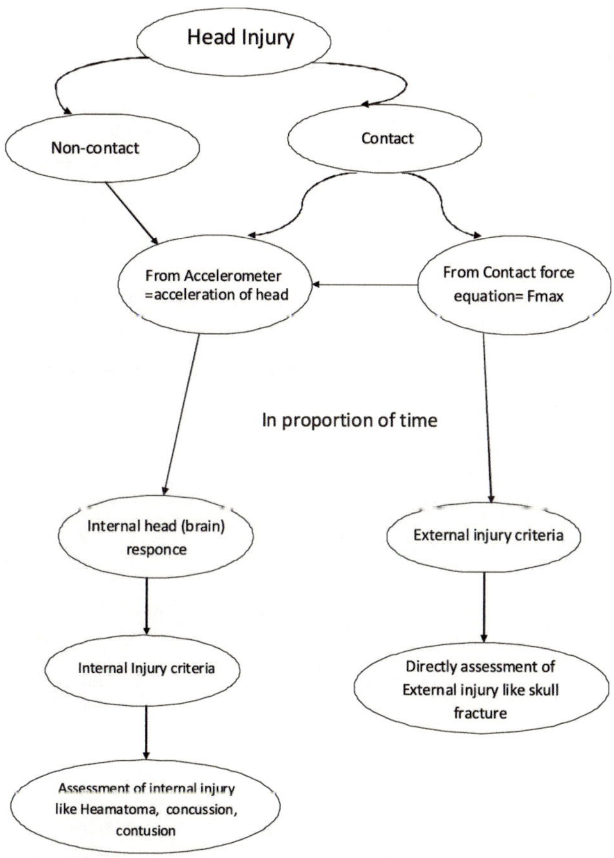

Figure 3. Flowchart of TBI injury assessment criteria development

The injury to the brain may be categorized in terms of, 1) The cause of injury, either contact vs. non-contact, 2) The type of injury, either primary in which the injury occurs at the time of initial injury producing event, or secondary where the injury results from some injury producing event but does not develop until somewhat later (through an intermediate process such as a metabolic effect), and 3) the type of injury, either focal (i.e. fairly localized) or diffuse (rather distributed) as shown in Figure 3.

In injury producing events, there are generally 3 collisions which occur [2]:

1. The "first collision" is where injury producing event occurs, e.g. the vehicle strikes another car or object and as a result the vehicle is rapidly decelerated and/or rotated.
2. The "second collision" is the movement of the occupants in the vehicle and their subsequent contact with the vehicle interior.
3. The "third collision" is when the internal organs of the occupant collide and/or move within the occupant.

2.1. Parameters that control head injury

A number of publications [17-43] discuss modeling and analyses of TBI using specific tools [44-45]. Gong [16] recently proposed a simple head-striker model to simulate the contact between a human head and a foreign-object striker. Based on the head-striker model, they formulated a contact force function, which is a function of time, impact mass, contact stiffness, impact velocity, and material properties of the head and neck. The contact force function was used for the estimation of the contact force between the human head and the foreign striker [5, 16].

2.1.1. Force

The contact force can be approximated [16] from the equation below. Then the estimated contact force may be used in two ways: 1) for the assessment of the exterior head injury, such as scalp damage, skull fracture, and 2) as input to the head model to predict the inner head injury, such as hematoma and brain injury [5].

$$Fmax = \frac{R*^{\frac{1}{5}} E*^{\frac{2}{5}} m*^{\frac{3}{5}} \Delta v^{\frac{6}{5}} E_{sh}^{\frac{1}{2}} h}{\left(\frac{1}{\sqrt{2.3}}\right) R*^{\frac{1}{5}} E*^{\frac{2}{5}} m*^{\frac{1}{10}} \Delta v^{\frac{1}{5}} R_{sh}^{\frac{1}{2}} (1-v_{sh}^{2})^{\frac{1}{4}} + \left(\frac{\sqrt{3}}{2}\right)\left(\frac{16}{15}\right)^{\frac{1}{10}} E_{sh}^{\frac{1}{2}} h}$$

The terminologies are explained in the original reference [16].

$$\frac{1}{R^{*}} = \frac{1}{R_{sol}} + \frac{1}{R_{sh}}, \quad \frac{1}{m^{*}} = \frac{1}{m_{sol}} + \frac{1}{m_{sh}} \quad and \quad \frac{1}{E^{*}} = 1 - \frac{v_{sol}^{2}}{E_{sol}} + 1 - \frac{v_{sh}^{2}}{E_{sh}}$$

2.1.2. Time duration

An analytical model [17] was proposed the impact of a fluid-filled spherical shell of mass (m_{sh}), thickness (h) and outer radius (R_{sh}) with a solid homogeneous isotropic elastic sphere of mass (m_{sol}) and outer radius (R_{sol}) at a relative velocity(D_v) as shown in Figure 4.

$$Tp = \pi \sqrt{\frac{\frac{3}{4}\left(\frac{16}{15}\right)^{\frac{1}{5}} m^{*\frac{4}{5}}}{R^{*\frac{2}{5}} E^{*\frac{4}{5}} \Delta v^{\frac{2}{5}}} + \frac{m^{*} R_{sh} \sqrt{(1-v_{sh}^{2})}}{2.3\, E_{sh} h^{2}}}$$

The shell was assumed to be filled with an inviscid fluid of density (ρ_f) and Bulk modulus (B) [5]. The impact mass, contact stiffness, impact velocity, angular velocity, accelerations, young's modulus, poison's ratio, time duration, height of the head(projectile) and an impactor influence severity of external forces were incorporated in the model. Effect of impact mass, contact stiffness and impact velocity on pressure-time histories have been described in [6].

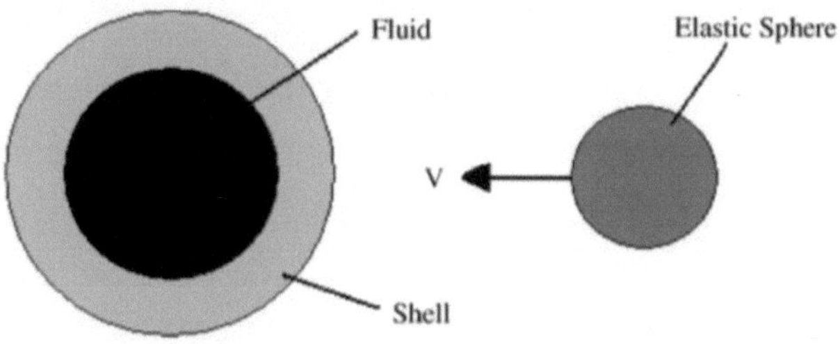

Figure 4. Illustrative representation of the analytical shell model.

(Rf- inviscid fluid of density, B- Bulk modulus, Esol; Esh and nsol; nsh are the Young's moduli and Poisson's ratio of solid and shell (Sphere), respectively. At occipital side of skull msh-mass of spherical shell-1.96 kg, h - thickness -0.00561, Rsh- radius of spherical shell-0.0725m, Msol- mass of solid, Rsol- outer radius of solid, Dv- velocity)[5]

2.1.3. Accelerations

The maximum acceleration of either projectile or head, assuming a quasi-static global response of the system, can be obtained by dividing the maximum force transmitted by the mass of the projectile or head, respectively [5]. For t>6ms impact time duration, neck force also needs to be considered, Figure 5, while calculating the resultant head accelerations. For short duration impacts (<6ms), the neck does not influence the kinematic head response [35, 43].

2.1.4. Contact area

Load or force to fracture/failure of the skulls of 12 unembalmed cadavers heads were reported by Yoganandan [19]. Using a hemispherical impactor with a 48 mm radius, they carried out impacts to various locations on the skull at a rate of 7.1–8.0 m/s. Failure loads ranged between 8.8 and 14.1 kN, with an average of 11.9 kN. Allsop [20] carried out temporo-parietal impacts on 31 unembalmed cadaver heads with two types of flat rigid impactors—one circular and 2.54 cm in diameter, the other a rectangular plate 5x10 cm. Fracture force for the small circular plate ranged between 2.5 and 10 kN with an average of 5.2 kN. Fracture force for the rectangular plate ranged between 5.8 and 17 kN with an

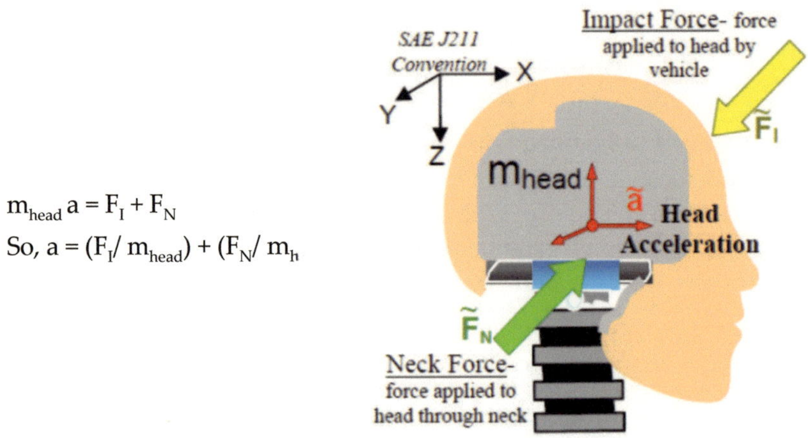

Figure 5. Force mechanism for head-neck [48]

average of 12.4 kN. The authors concluded that there is a significant relationship between contact area and fracture force. Thus, impacts with the ground are likely to require higher forces than with a smaller impactor [4].

2.2. Head injury criteria

Prior experiments on the capability of the human brain to hold impact forces were performed at Wayne State University using human cadavers and animal models [21, 22] as shown in Figure 6. This work led to the publication of the Wayne State Tolerance Curve [23, 24], a generally logarithmic curve that describes the relationship between the magnitude and duration of impact acceleration and the onset of skull fractures [7].

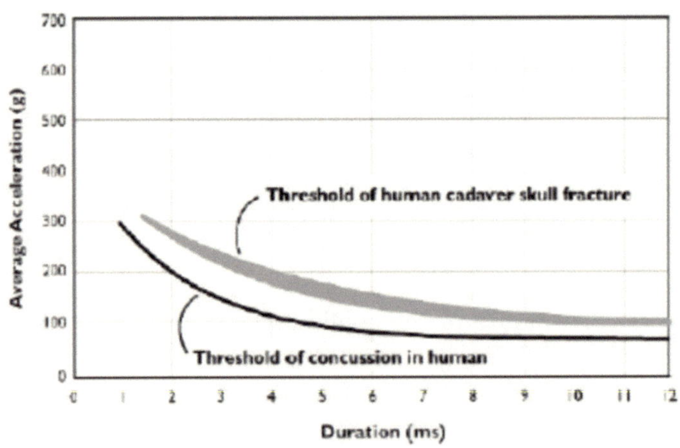

Figure 6. Wayne State Tolerance Curve [23, 24]

The relationship is nonlinear – the head can tolerate high accelerations for very brief periods but a longer exposure to a lower acceleration level may be damaging as well, Fig. 6. For a given degree of injury the logarithmic slope of the exposure time and acceleration graph is approximately –2.5. This relationship proposes the Severity Index (SI) as a measure of the injury potential of an impact [25]. SI is the integral of the acceleration time curve, weighted by the 2.5 factor observed in the Wayne State Tolerance Curve and calculated as

$$SI = \int_0^T a^{2.5}\, dt$$

Where *a (t)* is the acceleration-time pulse of the impact and *T* is its duration. An SI score of 1000 approximates the limit of human tolerance. Impacts with a higher score have a non-zero probability of causing a life-threatening brain trauma [7].

Severity Index SI [25] calculates distress of an impact in a way that quantifies the risk of head injury. In practice, SI scores are logical predictors of the injury potential of impacts that produce focal brain injuries. For impacts of lower intensity but longer duration, the SI calculation produces unreasonably high values that predict more severe injuries than those actually observed in cadaver experiments. The Head Injury Criterion (HIC) is an alternative measure of impact severity that is not subject to these errors. The HIC score is given by:

$$HIC = \max\left((t_1 - t_0) \left[\frac{1}{(t_1 - t_0)} \int_{t=t_0}^{t_1} a_t \, dt \right]^{2.5} \right)$$

Where t_0 and t_1 are the beginning and end times of the portion of the acceleration-time pulse being examined. The integral account for the duration of the acceleration and an iterative search found the time interval (t_0, t_1) to maximize the HIC score [7], Figure 7.

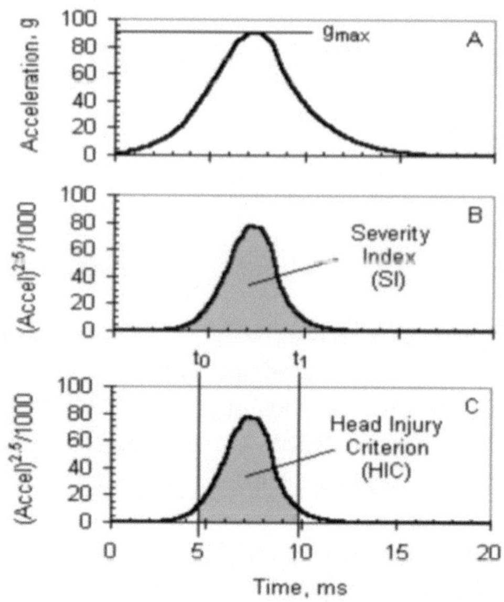

(A) Acceleration-time pulse from an impact between a surrogate head and an artificial turf surface, showing the peak value or gmax score.
(B) The same pulse with acceleration values
exponentiated to power 2.5. The SI score is the area under the curve
(C) As (B) but showing the time limits, t0 and t1, that maximize the HIC score.

Figure 7. Example of SI and HIC calculations [7]

A HIC score of 1000 represents the "safe" limit of human tolerance, above which the risk of a serious head injury is non-zero. In the sports surfacing world, HIC scores are the primarily determinant of playground surfacing, shock attenuation performance. Other terms of surfacing shock attenuation use a 200 g max limiting performance criterion, on that basis it approximates the HIC limit [7].

Empirically determined relationships between HIC scores and the probability of head injury [26, 27] are widely used in the automotive industry to estimate the risk of injury. Figure 8 shows examples of Expanded Prasad-Mertz Curves. Each curve estimates the possibility that an impact with a given HIC score will result in a specified level of head trauma [7]. Figure 8, also shows the relationship between the HIC score of a head impact and the probability of an injury.

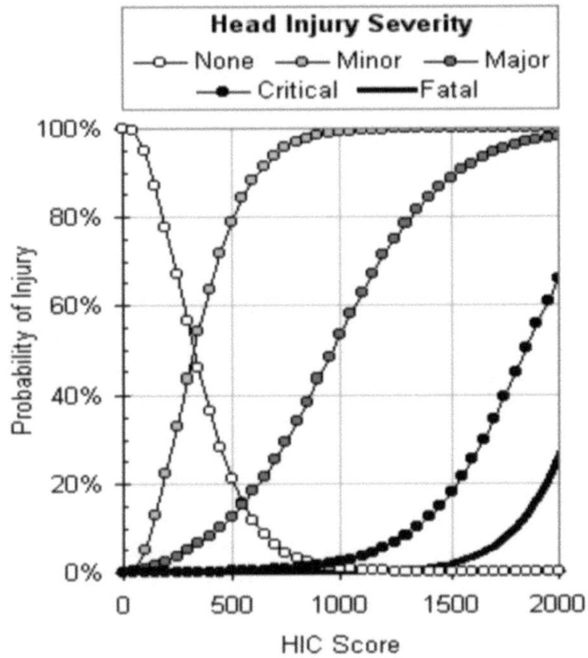

Figure 8. Expanded Prasad-Mertz Curves [7]

Computational simulation of real life head injury accidents has been used for various purposes. Some have compared AIS (abbreviated injury scale) scores for real life injuries to HIC scores or other indices of injury calculated from the reconstruction [4, 30-32]. HIC and tolerance levels have been explained [46-47] and tabulated in Table 1. Also, moderate and severe neurological injuries can only be distinguished with a criterion that is computed using intracranial variables and not with the sole global head accelerations [1]. More recently, there has been a move away from this approach of looking for a parameter that correlates well with overall severity of injury, and many are now focusing on determining tolerance limits of the head to specific lesion types, for example, acute subdural hematoma (ASDH), diffuse axonal injury (DAI) or skull fracture [4].

Head Injury Criteria	AIS Code	Level Of Brain Concussion And Head Injury
135 – 519	1	Headache or dizziness
520 – 899	2	Unconscious less than 1 hour – linear fracture
900 – 1254	3	Unconscious 1 – 6 hours – depressed fracture
1255 – 1574	4	Unconscious 6 – 24 hours – open fracture
1575 – 1859	5	Unconscious greater than 25 hours – large haematoma
> 1860	6	Non survivable

Table 1. Levels of Consciousness In Relation To Head Injury Criteria [46]

2.2.1. Injury criteria for Subarachnoid haematoma, contusion and skull fracture

Tolerance curves for ASDH due to rupture of bridging veins were experimentally produced in monkeys [28] and compared with human clinical data. It was concluded that bridging veins are highly sensitive to strain-rate and tend to rupture during impacts associated with high rates of increasing acceleration. As the duration of the pulse increases, higher levels of angular acceleration will be required in order to maintain the high strain rate necessary for rupture of bridging veins. Figure 9 shows tolerance curves for rhesus monkeys. For humans, a fall resulting in head acceleration of over 200 g and pulse duration of 3.5 ms or less would create conditions necessary for the production of bridging vein ASDH [4].

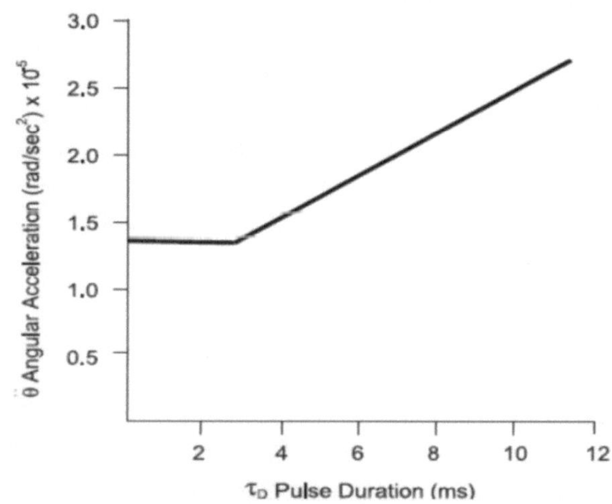

Figure 9. Injury tolerance curve for ASDH in rhesus monkeys [4]

Figure 10 shows the tolerance curves [29], derived for 5% critical strain, below which there is no axonal injury, and 10% critical strain, below which mild injury such as concussion could be expected and above which DAI can be expected. For impacts with very stiff contacts and short durations, the brain will move relative to the skull at impact, and thus a change in angular velocity of the skull will be of prime importance and causation of injury, Fig. 10. However, for impacts with softer structures, the brain will tend to move with the head, and will thus be subjected to the same accelerations [4].

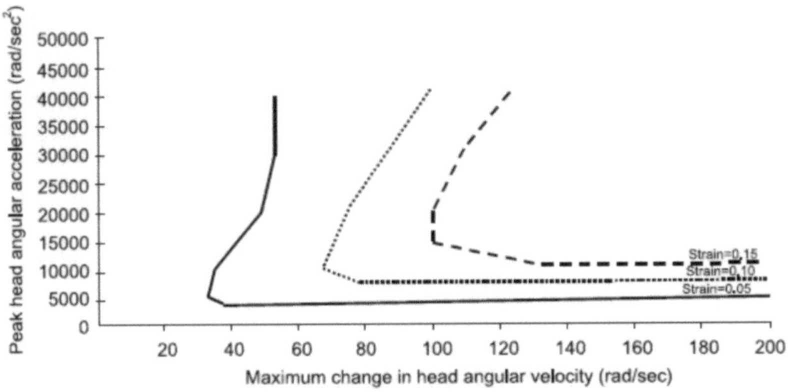

Figure 10. Tolerance curves for DAI and cerebral concussion [4, 29]

Tolerance of the head to skull fracture is much easier to determine than tolerance to intracranial injury. This is because of the definite relationship between force applied to the skull, and failure of cranial bone. Applied maximum force can be calculated from the equations discussed in the background section of head injury. Also from the Wayne State Tolerance Curve [23-24], tolerance of the head to skull fracture can be determined [4].

Tolerance limits to specific types of head injury were from reconstructing accidents and comparing the injuries sustained with parameters calculated from the reconstructions. For example, Auer [33] reconstructed 25 fatal pedestrian accidents using various methods, including computer simulations. Head acceleration and impact duration were calculated, and from these, the upper tolerance limit (lowest level of loading above which the specific injury is always observed) and the lower tolerance limit (highest value below which the injury never occurs) for various kinds of brain injury were determined, shown in Figure 11 [4].

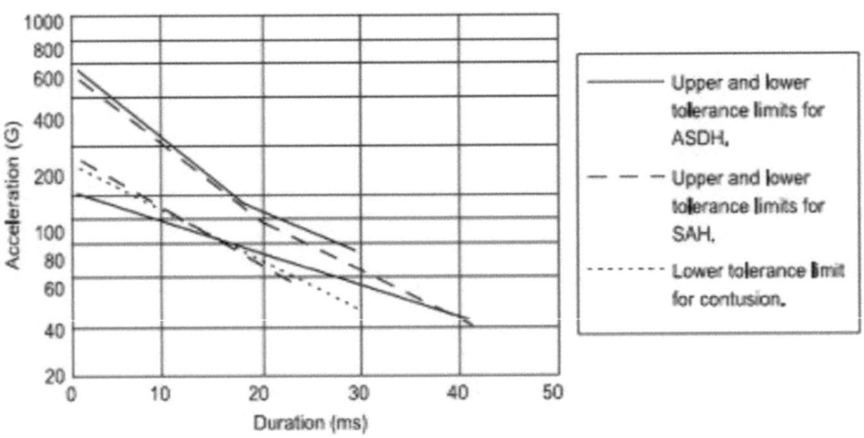

Figure 11. Upper and lower tolerance curves for ASDH, subarachnoid haematoma, and contusion [4]

The types of lesions examined were subdural haematoma, subarachnoid haematoma and brain contusions. While the authors did not elaborate on the relationship between the mechanical parameters and the lesions observed, they concluded that reconstructing pedestrian accidents could be a useful means of estimating tolerance limits for discrete brain injuries. However, due to lack of certainty about input variables, these are still very approximate estimations [4].

Since all head injury criteria are generally explained in terms of the resultant head accelerations, the resultant head acceleration are determined by placing an accelerometer to the desired points. The impact mechanisms are related with stresses, strain and pressure induced by the impact in the head which produce injury. Besides all these parameters affecting TBI with the resultant head accelerations, the following human head injury mechanisms and tolerance limits (stress, strain and pressure) were derived from accidents reconstruction [36-37].

A brain pressure reaching 200 kPa is an indicator for brain contusions, oedema and hematoma.
A brain Von Mises stress reaching 18 kPa is an indicator for moderate neurological injuries.
A brain Von Mises stress reaching 38 kPa is an indicator for severe neurological injuries.
A global strain energy of the brain skull interface reaching 5.4 J is an indicator for subdural hematoma and subarachnoidal bleeding.
A global strain energy of the skull reaching 2.2 J is an indicator for skull fractures.

3. Computational model validation

Finite element modeling and simulation of the human head biomechanics remain scarce in the literature. Only models that exist in the literature were reported by Ruan [18], and Willinger [35] and validated with limited experimental data. As FEM of the head finds wider applications in a diversity of fields, experimental validation is a critical key element [3]. Therefore, one of the objectives of this paper was to construct a 3D model of the head from Magnetic Resonance Imaging and validate FE analysis with available experimental data on stress induced by frontal and lateral impacts. Two sets of experimental data were used, from Nahum [42] and [34].

Previous research used various computational software: ULP models, ScanFE/RP (Simpleware Ltd.), FEA packages MSC/PATRAN; MSC/DYTRAN; ABAQUS; LS-DYNA3D (LS-DYNA3D, LSTC), MADYMO (Mathematical Dynamic Models) - may combine both multibody and FEM techniques, Test dummy- human body models to reconstruct the accident especially vehicle/car crash, Vtk and SUDAAN (based on CT scan sets). Methodology used in this paper is discussed below.

3.1. Methodology

3.1.1. *Software*

MIMICS software used in this study allows user to process and edit 2D image data (CT, μCT, MRI, etc.) to construct 3D models with accuracy, flexibility and user-friendliness, Figure 12. Besides smoothening, FEA, wide variety of boolean functions, the powerful segmentation tools allow user to segment the medical CT/MRI images, and take measurements. The designs can be modified based on the simulation outcomes and can be exported to the FEA/CFD packages [44]. Additional steps like assigning material properties, part sections, assemblies, load, boundary conditions and analysis for head models then exported into the ABAQUS [45].

3.1.2. FE Model properties

After exporting all four models in to ABAQUS, further simulation was done on randomly selected one of the four models. Tables 2-3 provide the subject specific dimensions and mechanical properties of the cadaver heads (computational models) used in this study. All four meshed-head models after exporting into the ABAQUS are shown in Figure 13. Further smoothening to reduce distorted elements was performed.

Skull: Total no of nodes: 51988 and total no. of elements: 210938

Brain: Total no of nodes: 36585 and total no. of elements: 145151

Young modulus, poisson's ratio and density are described in section 2.2

Material	Young modulus E(Mpa)	Poisson's ratio (ν)	Density ϱ (kg/m³)
Skull			
Outer table	7300	0.22	3000
Dipole	3400	0.22	1744
Inner table	7300	0.22	3000
CSF	2.19	0.489	1040
Brain	2190	0.4996	1040

Table 2. Young modulus, density and poison's ratio of the head [3]

	Skull 1	Skull 2	Skull 3	Skull 4
Volume (mm3)	663796.71	616368.55	687490.09	585843.68
Surface(mm2)	274159.51	263806.94	267584.24	249803.79
Thickness(mm)	11.2	10.36	12.18	8.64
Triangles	286772	268898	282116	238052

Table 3. Volumes, surfaces and number of elements (triangles) of four skulls

Elastic properties were assigned to brain. As per the [3, 35] viscoelastic or elastic properties do not make any fundamental change to the FEM response. Boundary condition details were provided in Figure 14.

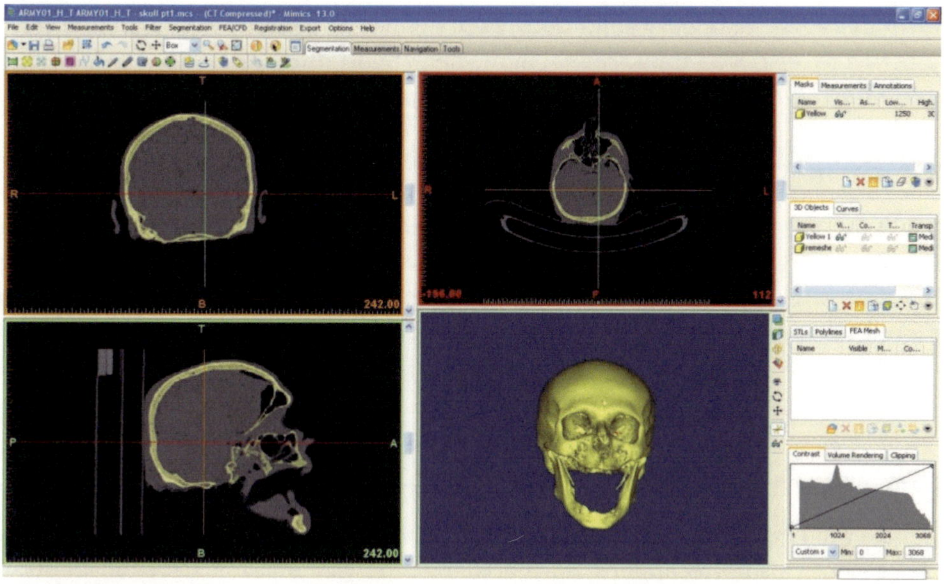

Figure 12. Use of Mimics to create 3D models of human head

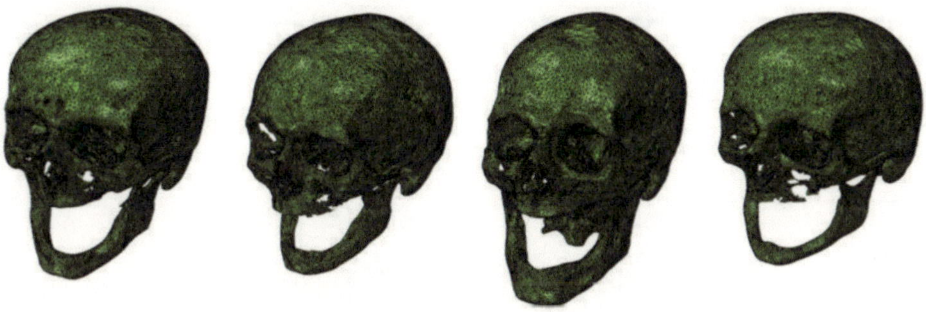

Figure 13. Four meshed skulls in ABAQUS

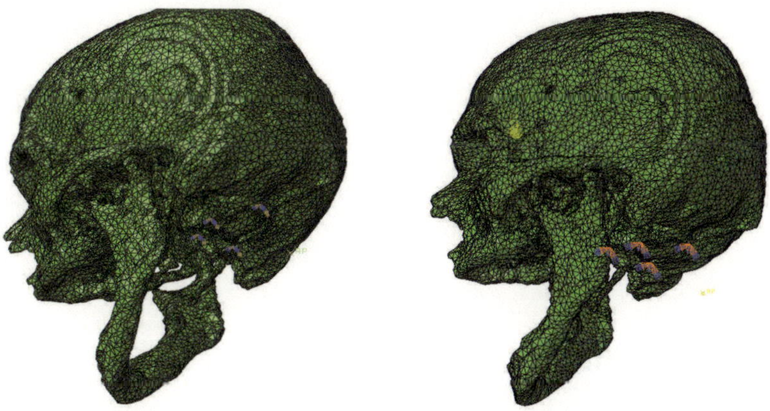

Figure 14. Blue/purple markers show the spots on at neck-head junction where boundary conditions were applied

As shown in the fig., boundary conditions were defined at the four points around the head-neck junction to restrict all transactional movement. Short duration impacts (<6ms), the neck does not influence the kinematic head response [35].

3.2. Validation

Validation of the model with experimental data was carried out while keeping the properties and load applications same. In order to reproduce the impact conditions, ~8000kN load was applied to the frontal side of the head, same as in Nahum's experiment [42]. Figure 15 shows pulse duration was kept 2 ms to reduce the time step cycles. Also, to compare the results for skull fracture with the prior experimental data [34], 8kN-16kN loads were applied.

3.2.1. Simulation versus experiment

To simulate the lateral impact, except the impact side on the head, all the other parameters were kept same, load was applied on the lateral side (left side) of the head as shown in Figure 14.

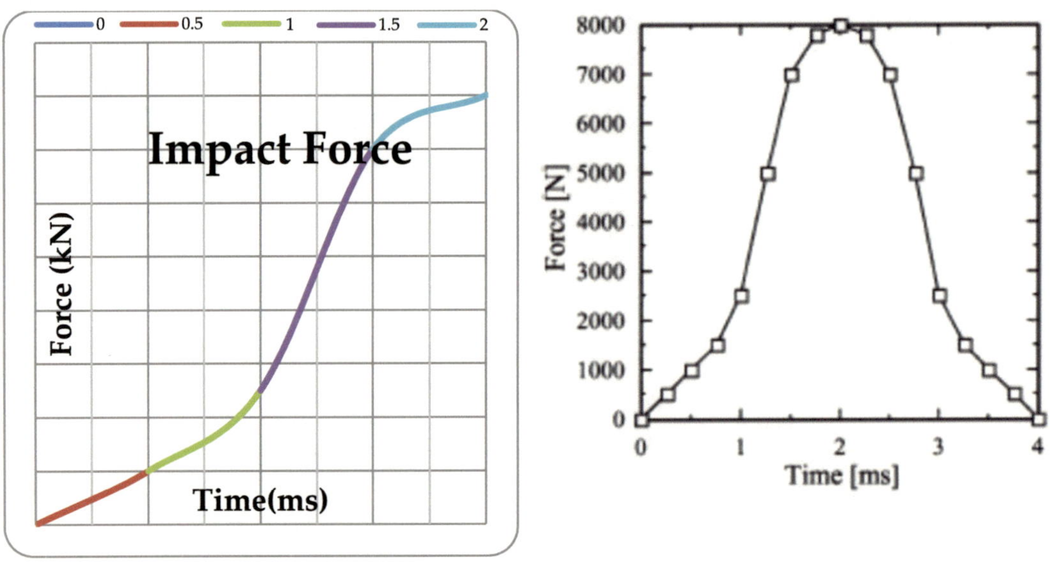

Nahum Experiment [42, 2] This simulation

Figure 15. Comparison of impact force- time curve between Nahum's experiment and current simulation

4. Results and discussion

The frontal impact on the head model predicted the same pressure on coup side as predicted in Nahum's Experiment [42]. This result validates the calibration runs as shown in Figure 16. The model duplicated the experimental response reasonably well, the only minor differences attributable to one or more of the following factors: the mesh fineness, reduced frame time steps, or by the material properties. An autopsy did not reveal any visible injury as a result of the Nahum experimental test and, therefore, based on this observation the brain tolerance thresholds were: compression: 234 kPa, tension: 186 kPa.

A 16kN load applied to the frontal side of the head while other parameters kept same. Analysis ran 1.1E-3 seconds due to large number of damaged volumes created after that instance. This was consistent with the Yoganandan [19] and Allsop [20] that fracture occurs because of applied force range of 8.8-17 kN. The intracranial pressure reached 200 kPa which was an indicator for brain contusion, oedema, and haematoma, but the pressure exceeded 200 kPa and reached 249 kPa, which was only slightly higher than the threshold limit of brain (234kPa), see Figure 17.

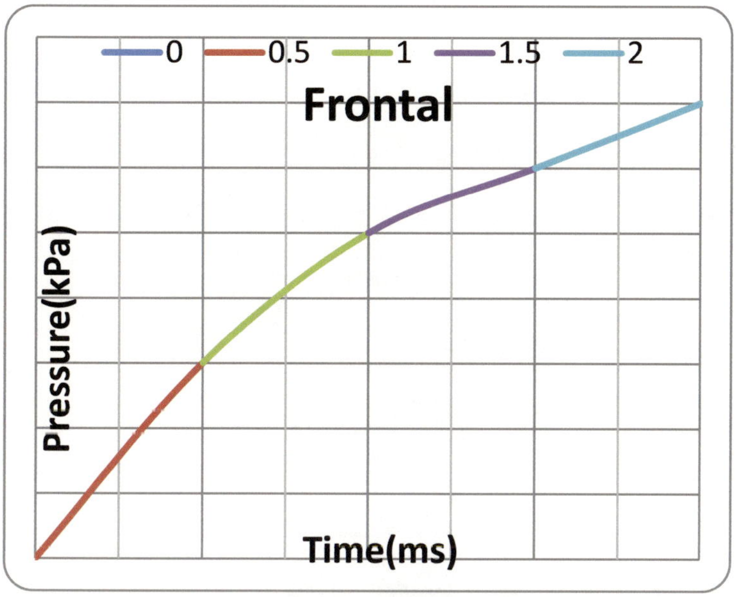

Figure 16. Frontal pressure- time curve results for comparison with Nahum's experimental results.

The history output of strain energy of the model also seemed to be at 2.2 J consistent with indications of skull fracture Figure 17. Also, from Newton's second law, the resultant acceleration of head can be calculated as a=16kN/4.5kg (sample of patients were of male adults in the age range of 30-50 and the mass of head was considered nearly ~ 4.5 kg).

A fall resulting in head acceleration of over 200 g and pulse duration of 3.5 ms or less would create conditions necessary for the production of bridging vein ASDH [4, 28]. Also, a= 355.5g is > 150g represents the HIC > 2000 which is non-survival head injury. Thus, these results depict that the model is valid for the further analysis in injury biomechanics.

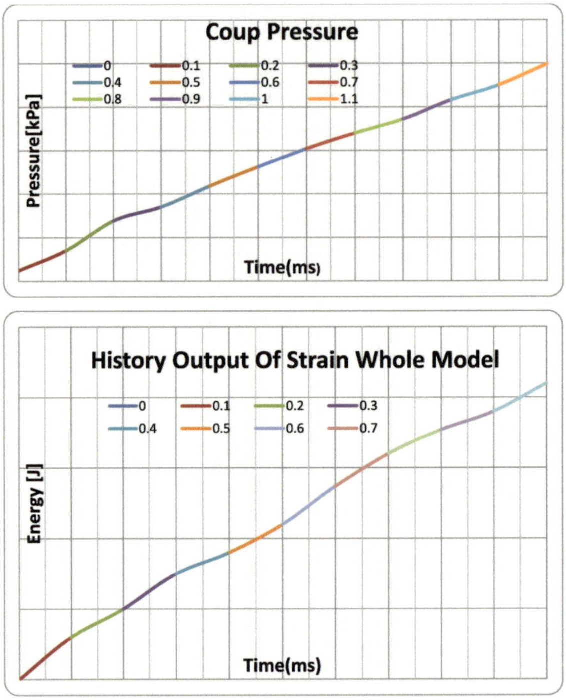

Figure 17. Frontal pressure- time curve and history output of whole strain model after applying 16kN

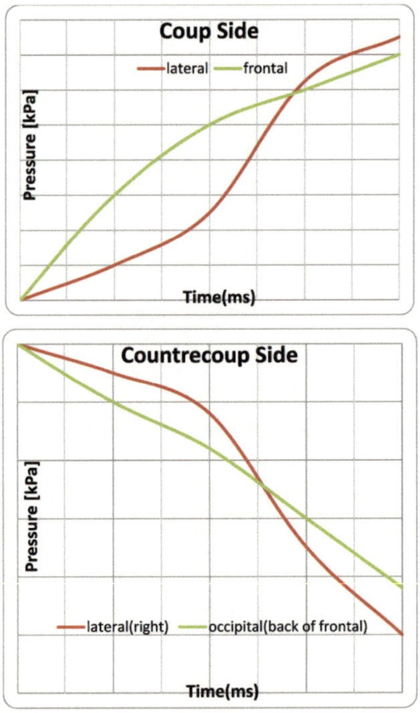

Figure 18. Comparison of pressure-time curves at coup and countercoup sides between lateral and frontal impact

Simulation result shows that the relative risk and severity of TBI in lateral impacts are higher than in the frontal impacts. Figure 18 shows the pressure-time history for coup and countercoup (at and opposite side of the impact, respectively) sides of the model. It shows quite similar pressure-time curve compared to frontal one. However, the lateral impact produces 6.67% more pressure at coup side as compared to frontal impact. The results of countercoup side support the prior analysis predicting only 14% higher tensile stress by lateral as compared to frontal impacts.

Statistical analysis carried out on 1115 occupants who were the victims of lateral and non-lateral automobile impacts [40-41], TBI occurred from lateral impacts were more severe than those resulting from non-lateral impacts.

5. Conclusion

The paper reviewed the head injury mechanisms and criteria. A computational framework was developed to biomechanical parameters to assess the injury, and validate the finite element models of the human head. The comparison of the stress/pressure incurred by lateral and frontal impacts in the coup and countercoup side of the head was presented. The model has been validated against the two sets of experimental results: one obtained in frontal impact and the other using head tolerance/skull fracture data.

Although the results obtained from the study involved a degree of inaccurateness (i.e., model had around 6500 distorted elements, 3 layers of skull was assigned as a one layer having the mechanical property (young's modulus, poison's ratio and density) as an average of those 3 layers), they do nonetheless confirm that through proper sets of MRI data, analytical modeling is applicable in injury biomechanics.

It is concluded that the lateral impacts are more severe than the frontal impacts. Therefore, it is imperative that victims of lateral impacts are at more risk for TBI than the frontal impacts. This information may be useful in injury assessment and developing sensors to alleviate lateral impacts to prevent traumatic brain injuries.

Author details

Aalap Patel and Tarun Goswami*

Department of Biomedical, Industrial and Human Factors Engineering, Wright State University, Dayton, OH, USA

6. References

[1] Marjoux, D., Baumgartner, D., Deck, C., and Willinger, R., Head injury prediction capability of the HIC, HIP, SIMon and ULP criteria –New injury criteria for the head. Accident Analysis and Prevention 40(3), 1135-1148, 2008.

* Corresponding Author

[2] Gilchrist, M. and O'Donoghue, D., 2000, Simulation of the Development of Frontal Head Impact Injury, *Comp. Mech.*, 26, 229-235.

[3] Zong*, Z, H P Lee and C Lu, "A three-dimensional human head finite element model and power flow in a human head subject to impact loading". JOURNAL OF BIOMECHANICS, 39, no. 1 (2006): 284-292. (United Kingdom).

[4] O'Riordain K, Thomas PM, Phillips JP, Gilchrist MD. Reconstruction of real world head injury accidents resulting from falls using multibody dynamics. Clin Biomech 2003; 18:590–600.

[5] E. A. C. Johnson and P. G. Young, "On the use of a patient-specific rapid-prototyped model to simulate the response of the human head to impact and comparison with analytical and finite element models," *J. Biomechan.*, vol. 38, pp. 39–45, 2005.

[6] Shi Wei Gong, Heow Pueh Lee, Chun Lu., Dynamic response of a human head to a foreign object impact, Inst. of High Performance Comput., Singapore;. Biomedical Engineering, IEEE Transactions on. 04/2008; 55(3):1226-1229

[7] Shorten M.R. & Himmelsbach, J.A. (2003),Sports surfaces and the risk of traumatic brain injury, pp 49-69 in sports surfaces(Eds. B.M. Nigg, G.K. Cole, D.J. Stefanyshyn)Calgary, University of Calgary

[8] http://www.pas.rochester.edu/~blackman/ur10helmetsimpact.pdf

[9] Portions from Pike, J. A., Automotive Safety: Anatomy, Injury, Testing and Regulation, SAE, 1990

[10] http://homepages.nyu.edu/~eh597/Meninges.htm

[11] "Head Injury Criterion and the ATB", Presented at the 2004 ATB Users' Conference, Salt Lake City, UT (http://www.mchenrysoftware.com/HIC%20and%20the%20ATB.pdf)

[12] Copyright © 2001 - 2011 BrainInjury.com / Steven (Woody) Igou

[13] http://www.peoplecentric.info/tbi-autism-aspeger-assocations/

[14] Ommaya, A.K., Thibault, L., Bandak, F.A., 1994. Mechanisms of impact head injury. Int. J. Impact Eng. 15, 535–560.

[15] Jennett, B., 1996. Epidemiology of head injury. J. Neurol. Neurosurg. Psychiatry 60, 362–369.

[16] S. W. Gong, H. P. H. P. Lee, and C. Lu, "An approach for the estimation of contact force on a human head induced by a foreign-object impact," *IEEE Trans. Biomed. Eng.*, vol. 54, no. 5, pp. 956–958, May 2007.

[17] Young, P.G., 2003. An analytical model to predict the response of fluid-filled shells to impact—a model for blunt head impacts. Journal of Sound and Vibration 267, 1107–1126.

[18] J. S. Ruan, T. B. Khalil, and A. I. King, "Dynamic response of the human head to impact by three-dimensional finite element analysis," *J. Biomechan. Eng., vol. 116, pp. 44–50, 1994.*

[19] Yoganandan, N., Pintar, F.A., Sances Jr., A., Walsh, P.R., Ewing, C.L., Thomas, D.J., Snyder, R.G., 1995. Biomechanics of skull fractures. J. Neurotrauma 12, 659–668.

[20] Allsop, D.L., Perl, T.R., Warner, C.Y., 1991. Force/deflection and fracture characteristics of the temporo-parietal region of the human head. In: Proc. 35th Stapp Car Crash Conf., pp. 269–278.

[21] Gurdjian ES, Webster JE, 1945. Linear acceleration causing shear in the brainstem in trauma of the central nervous system. Mental Adv Dis 24:28.

[22] Gurdjian ES, Webster JE, Lissner HR, 1955. Observations on the mechanism of brain concussion, contusion and laceration. Surg Gynec Obstet 101:680-690.

[23] Lissner HR, Lebow, M, Evans FG, 1960. Experimental studies on the relation between acceleration and intracranial changes in man. Surg Gynecol Obstet 11: 329-338.

[24] Patrick LM, Lissner HR, Gurdjian ES, 1963. Survival by design – head protection. Proc 7th Stapp Car Crash Conference 36: 483-499.

[25] Gadd CW, 1966. Use of a weighted impulse criterion for estimating injury hazard. Proc 10th Stapp Car Crash Conference; SAE Paper 660793, Society of Automotive Engineers, Warrendale PA, USA.

[26] National Highway Traffic Safety Administration (NHTSA), Department of Transportation, 1997.FMVSS201, Head Impact Protection, 49 CFR §571.201.

[27] Prasad P, Mertz HJ, 1985. The position of the United States delegation to the ISO working group on the use of HIC in the automotive environment. SAE Paper# 851246 Society of Automotive Engineers, Warrendale PA, USA.

[28] Gennarelli, T.A., Thibault, L.E., 1982. Biomechanics of acute subdural hematoma. J. Trauma 22, 680–686.

[29] Ryan, G.A., Vilenius, A.T.S., 1994. Field and analytic observations of impact brain injury in fatally injured pedestrians. In: Proceedings of the Head Injury_94 Conference, Washington, DC sponsored by US National Highway Traffic Safety Administration and George Washington University, 12–14 October, pp. 181–188.

[30] Enouen, S.W., 1986. Development of experimental head impact procedures for simulating pedestrian head injury. In: Proc. 30th Stapp Car Crash Conf., pp. 199–218.

[31] MacLaughlin, T.F., Wiechel, J.F., Guenther, D.A., 1993. Head impact reconstruction—HIC validation and pedestrian injury risk. In: Proc. of the SAE Conf. on Accident Reconstruction: Technology and Animation III, pp. 175–183

[32] Mohan, D., Bowman, B.M., Snyder, R.G., Foust, D.R., 1979. A biomechanical analysis of head impact injuries to children. J. Biomech. Eng. 101, 250–260.

[33] Auer, C., Schonpflug, M., Beier, G., Eisenmenger, W., 2001. An analysis of brain injuries in real world pedestrian traffic accidents by computer simulation reconstruction. In: Proc. Int. Soc. Biomechanics XVIIIth Congress.

[34] Doorly MC, Gilchrist MD. The use of accident reconstruction for the analysis of traumatic brain injury due to head impacts arising from falls. *Computational Methods in Biomechanics and Biomedical Engineering*. 2006;9:371–377

[35] Remy Willinger, Ho-Sung Kang, and Baye Diaw, "Three-Dimensional Human Head Finite Element Model Validation against Two Experimental Impacts", Annals of Biomedical Engineering, Vol. 27, pp. 403-410, 1999.

[36] D. Baumgartner[a], R. Willinger[a], J.S. RAUL[b]Finite element modeling of the human head and application to forensic medicine, *19ème Congrès Français de Mécanique, Marseille, 24-28 août 2009*

[37] Baumgartner D., Willinger R., Human head tolerance limits to specific injury mechanisms inferred from real world accident numerical reconstruction, Revue Européenne des Eléments Finis, vol. 14, n° 4-5, pp. 421-444, 2005.

[38] Ward, C. C., M. Chan, and A. M. Nahum. Intracranial pressure—A brain injury criterion. In: Proceedings of the 24th Stapp Car Crash Conference, SAE Paper No. 801304, 1980, pp. 347–360.

[39] Bazarian JJ, Fisher SG, Flesher W, Lillis R, Knox KL, Pearson TA., Lateral automobile impacts and the risk of TBI, Ann Emerg Med. 2004 Aug;44(2):142-52.

[40] Morris A, Hassan A, Mackay M, et al. Head injuries in lateral impact collisions. Accid Anal Prev. 1995;27:749-756.

[41] McLellan BA, Rizoli SB, Brenneman FD, et al. Injury pattern and severity in lateral motor vehicle collisions: a Canadian experience. J Trauma. 1996;41:708-713

[42] Nahum, A. M., R. Smith, and C. C. Ward. Intracranial pressure dynamics during head impact. In: Proceedings of the 21st Stapp Car Crash Conference, 1977, pp. 339–366.

[43] Willinger, R., L. Taleb, and C. M. Kopp. Modal and temporal analysis of head mathematical models. *J. Neurotrauma* 12:N4, 1995.

[44] http://www.materialise.com/mimics

[45] ABAQUS Dynamic Explicit Version 6.9 Hibbitt, Karlsson& Sorenson,Inc., Pawtucket, RI

[46] http://www.eurailsafe.net/subsites/operas/HTML/appendix/Table13.htm

[47] http://www.eurailsafe.net/subsites/operas/HTML/Section3/Page3.3.1.4.htm

[48] Jason Kerrigan1, Carlos Arregui2, Jeff Crandall1,Pedestrian head impact dynamics-comparison of dummy and PHMS IN SMALL SEDAN AND LARGE SUV IMPACTS (www-nrd.nhtsa.dot.gov/pdf/esv/esv21/09-0127.pdf)

Chapter 7

Cervical Spinal Injuries and Risk Assessment

Mary E. Blackmore, Tarun Goswami and Carol Chancey

Additional information is available at the end of the chapter

http://dx.doi.org/10.5772/50588

1. Introduction

As of mid 2001, out of the 34000 spinal injuries that took place in the previous year, over half (55%) were cervical spine injuries [18]. Additionally, out of all the cervical spinal injuries incurred by patients in the United States every year, 15% of those injuries are fatal [7]. Motor vehicle crashes are the leading cause of death for persons under 45 years of age, and the number one cause of head and spinal cord injury [1]. This study compiles and analyzes data that may be used to assess risk of cervical injury.

The cervical spine is a very complex anatomical structure. Any neck injury can have debilitating, and sometimes life threatening consequences. Although spinal cord injuries vary significantly from the injuries of the vertebral column, they result from structural deformities and were therefore studied prior to this analysis [21]. For both spinal cord injuries and vertebral body fractures, motor vehicle accidents are the most common causes of neck injuries in both Canada and the United States (Figure 1). Out of the 1.4 million annual American spinal cord injuries, approximately 280,000 of those are motor vehicle induced. One out of every five drivers are involved in a traffic accident each year [1]. Figure 1 illustrates the most common mechanisms for cervical spine injuries. Another interesting aspect is the increase in violence, which in turn could impact the number of violence related spinal injuries and incidents (shooting, stabbing, etc.) [21].

Sports and leisure activities account for a significant amount of neck trauma. They can be broken down by both the activities most likely to cause injury, as well as the injuries accounted for in specific sports. Understanding what particular actions and motions within each activity actually contribute to the risk of injury, has helped improve sporting equipment and decrease the number of neck injuries associated with various sports. Table 1 lists the most common leisure activities associated with neck injuries [20]. Diving and surfing involve more injuries than football (Table 1). This is most probably because football has grown in popularity since 1989, when this data was originally compiled.

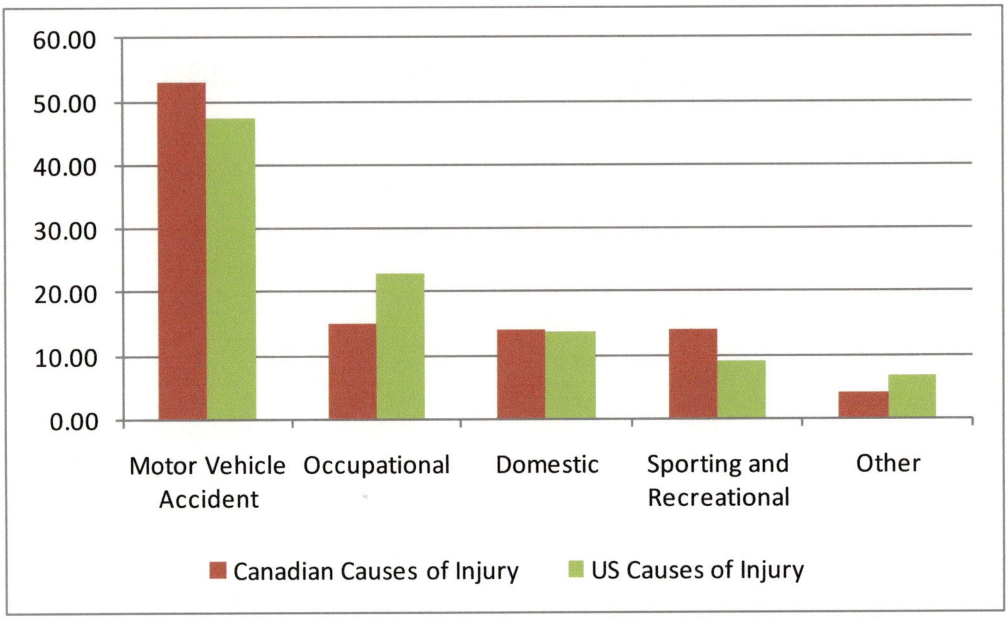

Figure 1. Frequency of Activities Causing Cervical Injuries, in Percentage [7]

From Table 1, it can be seen that many common everyday activities offer the potential for serious injury. In diving, fractures and dislocations are the most commonly seen. Not only are the diver's form and function important, but depth of water, angle of entry and head velocity also prove crucial to injury severity. Degrees of participation within the various sports also play a role in the frequency of injury. Today, more athletes participate in football, making the likelihood of injury higher than it was in 1989.

Cause of Injury	Number	Injuries (%)
Diving	82	54
Football	16	11
Gymnastics	5	3
Snow Skiing	5	3
Surfing	29	19
Track and Field	3	2
Trampoline	2	1
Water Skiing	7	5
Wrestling	3	2
Total	152	

Table 1. Sports Activities Causing Cervical Spinal Injuries

2. Research background: Analysis of cervical spinal tolerances and injuries

With motor vehicle accidents being the leading mechanism behind both spinal cord and vertebral fracture injuries, significant research has focused on improving the design and safety of automobiles [14]. Vertebral fractures can occur at any level to any degree, and can be caused by various types of loading. Figure 2 illustrates the frequency of injury to various levels of the cervical spine as well as the type of loading that causes that injury.

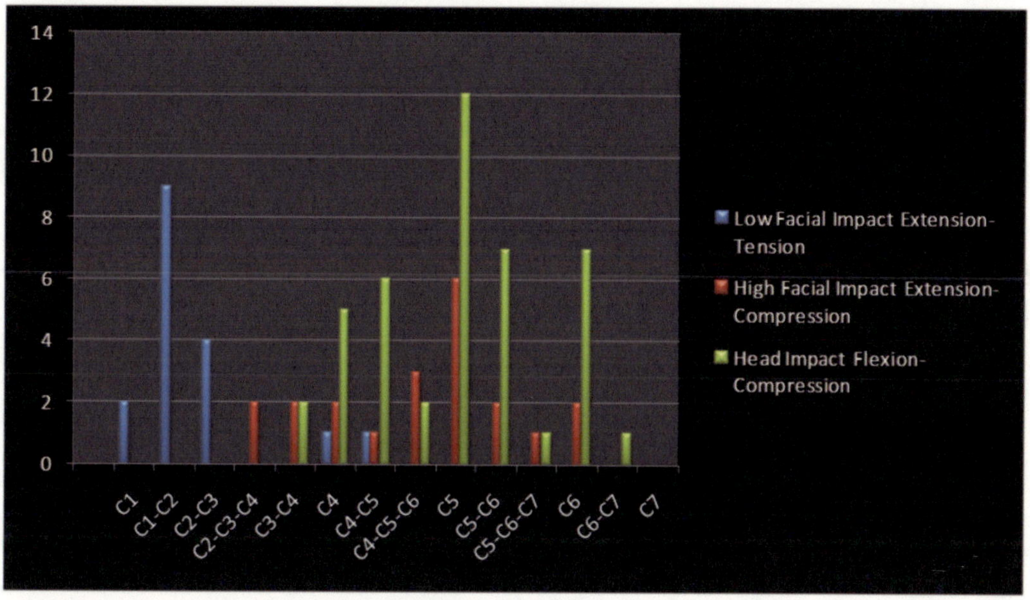

Figure 2. Frequency of Fractures per Level, Based on Type of Impact Loading

This data is also represented by use of the Abbreviated Injury Scale (AIS). Especially in the case of automobile and motorcycle accidents, the victim is usually injured because of some form of head contact with another object. Many factors contribute to the severity of the injury: position of the head and neck, the impact site, the nature of the impacted surface and the direction of the cervical spine loading. Determining the relationship between all of these variables is very complex and until recently has been based on frequency of occurrence data. Table 2 is a summary of the AIS scale characteristics. Figure 3 illustrates the amount of injuries that occurred in over 100 automobile accidents, in which a passenger attained a neck injury of some degree [25]. The degree of the injury is indicated by the AIS score.

With respect to frequency, the 5th and 6th vertebrae of the cervical spine see the most injury, and have the most critical injuries (AIS 5). This is the most significant score one can achieve and still survive, meaning that the ability to minimalize the chance of injury through improved vehicles and vehicle interior is crucial to assessing and preventing risk.

AIS Score	Injury
1	Minor
2	Moderate
3	Serious
4	Severe
5	Critical
6	Unsurvivable

Table 2. AIS Scoring Details [12]

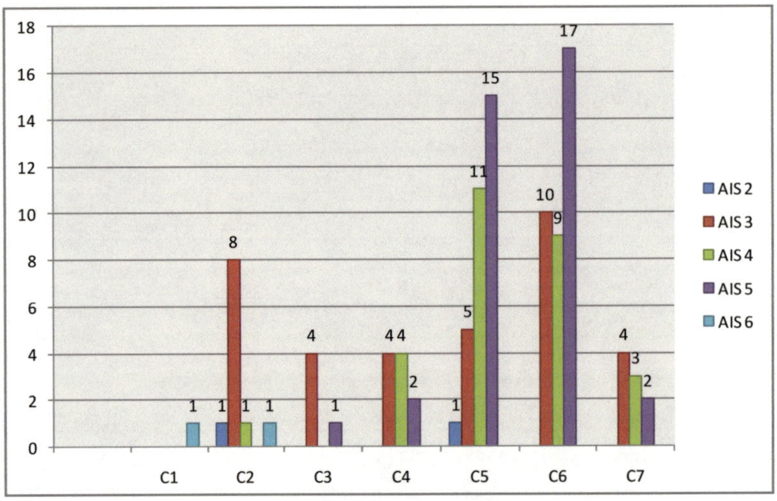

Figure 3. Automobile Injuries by Cervical Spinal Level and Their Associated AIS Scores

To assess the potential for injury, the loads that the cervical spine can withstand during various activities must first be understood. This can be broken down into the type of loading seen during the various activities. There are 5 main types of loading incurred by the cervical spine: Flexion-Extension, Compression, Tension, Torsion and Horizontal Shear. The main injury mechanisms are shown in Figure 4, while all are described in succeeding sections.

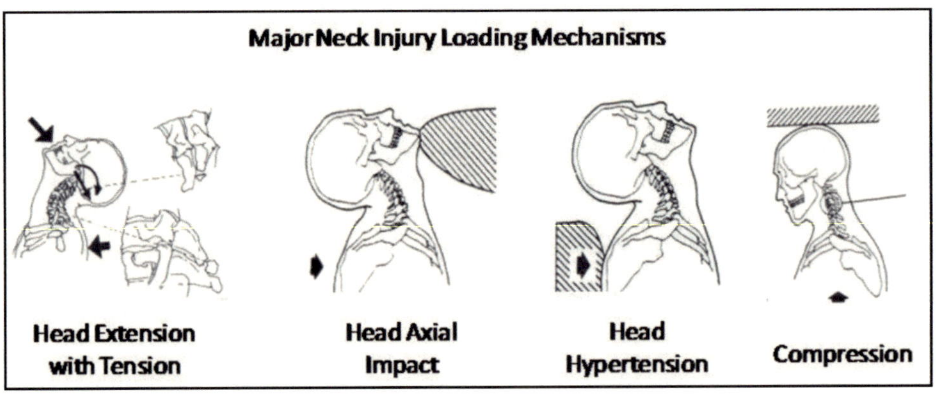

Figure 4. Injury Mechanisms of the Cervical Spine

2.1. Flexion-extension

A lower bound for the risk of injury of the neck under flexion-extension is based on the bending moment at the occipital condyle. For extension, noninjurious loading occurs at 35 ft-lbs (47.3 Nm), with ligamentous injury occurring above 42 ft-lbs (56.7 Nm). For flexion, pain can be felt at a load of 44 ft-lbs (59.4 Nm) and the risk of significant structural damage occurs around and above 140 ft-lbs (189 Nm). Bending moments of 24 Nm and resultant forces of 130 N can be felt without injury, but anything beyond that will most likely result in injury [10, 12, 15].

2.2. Compression

This is the most researched type of loading, as a large amount of data can be found on compressive load analysis until failure is reached. The type of axial loading and the degree of constraint imposed on the contacting surfaces causes the results to vary among investigators. It has however been proven that compression-flexion and compression-extension injuries occur under smaller axial loads than pure compressive injuries. Bilateral facet dislocations can occur at loads of 1720 ± 1230 N, with flexion injuries occurring at approximately 2000 N. When dealing with pure compression, injuries have been reported under loads of 4810 ± 1290 N of loading. The average peak head and neck loads that can be reached before structural injury occurs are 5.9 ± 3.0 kN and 1.7 ± 0.57 kN, respectively [10, 14].

2.3. Tension

Tension loading is not a commonly studied area of research. Past studies have indicated that the cervical spine has a tensile loading tolerance of 1135 N. With respect to automobile accidents, the average cranial accelerations are usually between 40 and 50 g [11]. This results in an estimated traction load of the Atlas (C1) of 1600 – 2000 N [3]. These types of loads produce disc damage, joint capsule tears and skull and vertebral fractures. The mean force at failure of intervertebral discs is 581 ± 220 N, but much still needs to be identified with respect to the amount of tensile force the vertebral bodies, and the entire cervical spine can withstand [19].

2.4. Torsion

The estimated lower bounds of axial torsional tolerance are between 13.6 ± 4.5 Nm and 17.2 ± 5.1 Nm [13]. This amount of torque produces upper cervical spinal injuries. It has also been estimated that the cervical spine can withstand approximately 114 ± 6.3^0 of rotation before injury occurs [6, 24].

2.5. Horizontal shear

Another area not too commonly researched is the amount of horizontal shear needed to produce cervical spinal injury. Most of this research is conducted to learn more about the mechanisms that cause occipito-atlantoaxial injuries. Transverse ligament rupture has been seen to occur at a load of 824 N, with anterior shear of the atlas [5]. Odontoid fractures reportedly occur at 1510 ± 420 N of shear force [4]. Additional, higher tolerances have been reported up to 5500 ± 2500 N when the shear force is applied at the chest [2].

3. Methods – Assessing the probability of injury

In this section graphical presentation of risk of injury presented. Envelops pertaining to risk free and high risk injury needs to be established. Based on the data compiled here, and by others [14] the succeeding curves have been developed to establish the amount of existing risk associated with various types of cervical spinal loading. The first set of curves in Figure 5 illustrate the tolerance of tensile loads of 5 different human body types, (mannequins were used to test and extract data for each of these body types), based on the amount of time the load is applied. The Hybrid III Family of Mannequins is a well established group used for testing the effects of various types of loading on different sized bodies. Within this family are five mannequins: (i) A 3 Year Old, (ii) a 6 Year Old, (iii) a Small Adult, (iv) a Midsize Adult and (v) a Large Adult. Table 3 displays the characteristics of each particular mannequin.

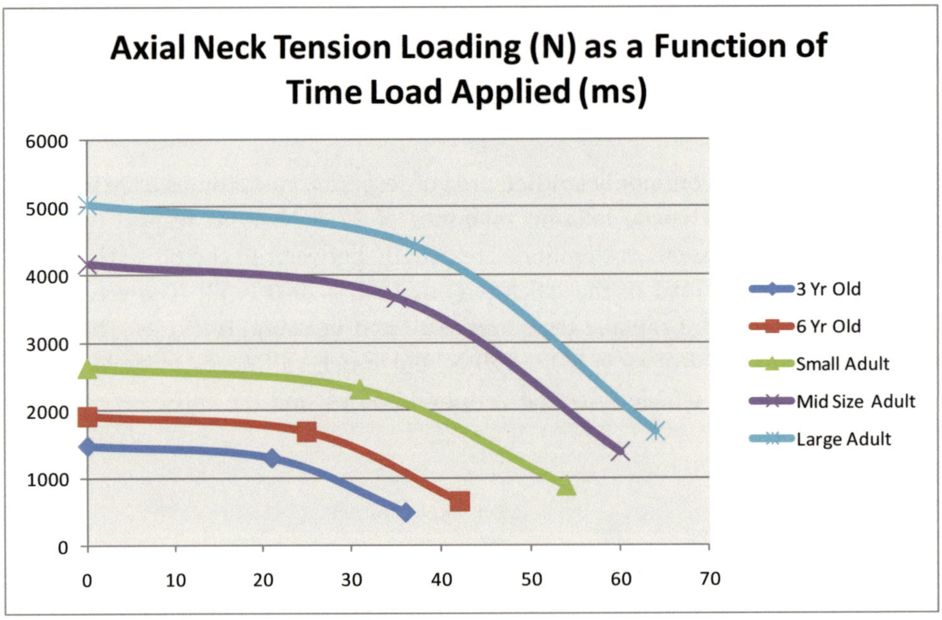

Figure 5. Tension Loading of Five Different Body Types Retrieved from Dummy Data [13]

No height is given for the 3 and 6 year old mannequins because their development was based largely in part on estimates and approximations. This remains a difficult parameter to analyze because so little data exists on the effects of accidents on children. Figure 5 displays the Axial Tension (Newtons) Tolerance of the mannequins with respect to time (milliseconds).

Anything above each individual curve in Figure 5 indicates the potential for significant neck injury due to tension loading, while anything below indicates that significant neck injury due to loading is highly unlikely. Compression and shear data was analyzed and compiled in a similar fashion, to develop the curves in Figures 6 and 7, respectively. Anything above each individual curve indicates the potential for significant neck injury due to compression loading, while anything below means that significant neck injury due to loading is highly unlikely.

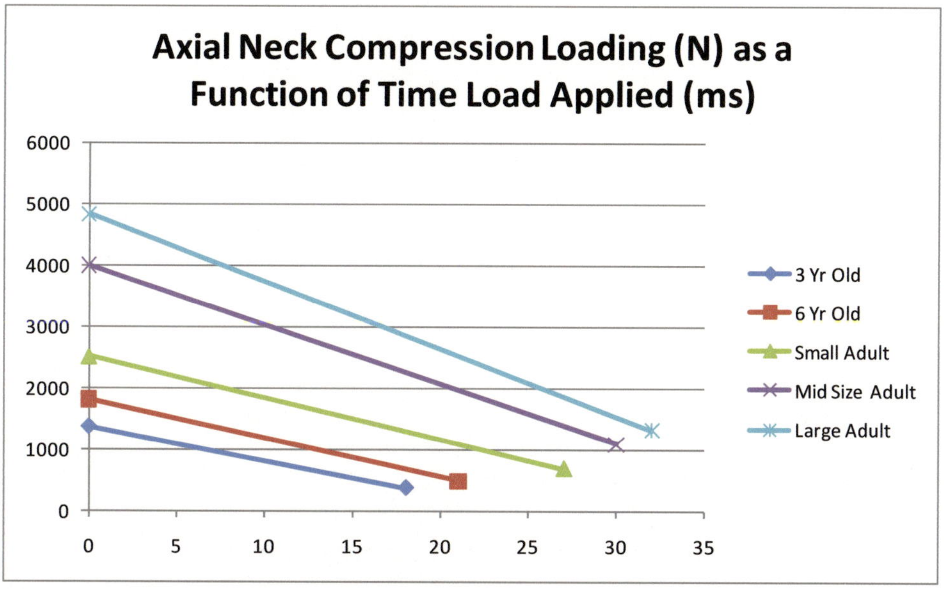

Figure 6. Compression Loading on the Neck [13]

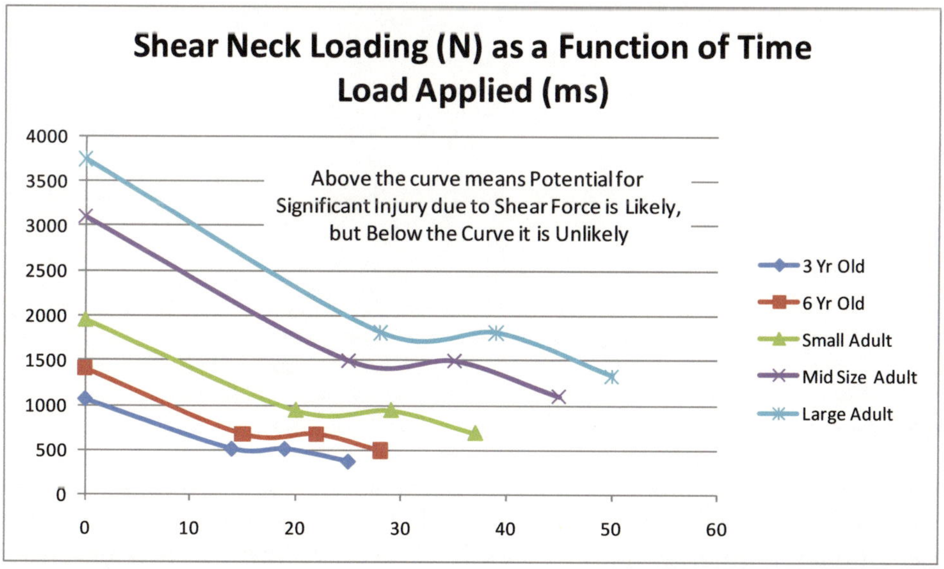

Figure 7. Shear Neck Loading Data [13]

Each of the curves for compression, tension and shear loading were analyzed. From these curves, specific data points were extrapolated and both standard deviations and standard error for each point were calculated. The extrapolated data for compression, tension and shear loading are illustrated in Figures 8, 9, and 10 respectively.

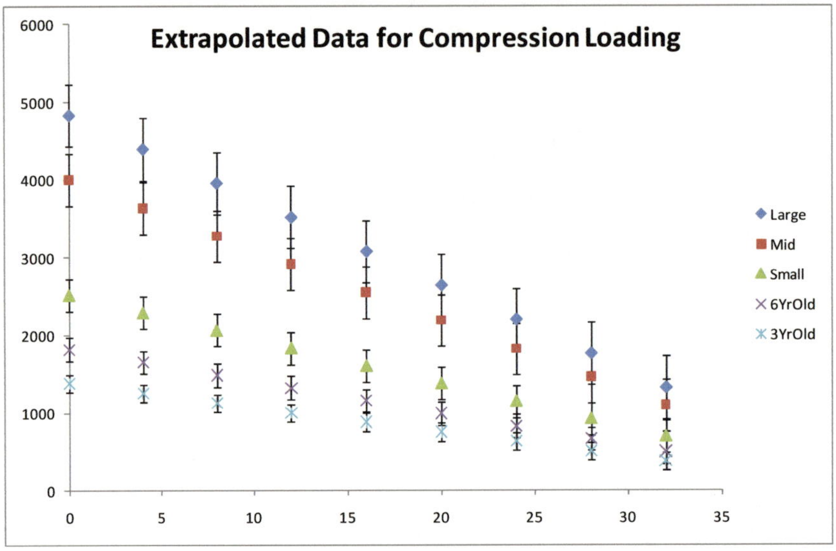

Figure 8. Extrapolated Compression Loading Data

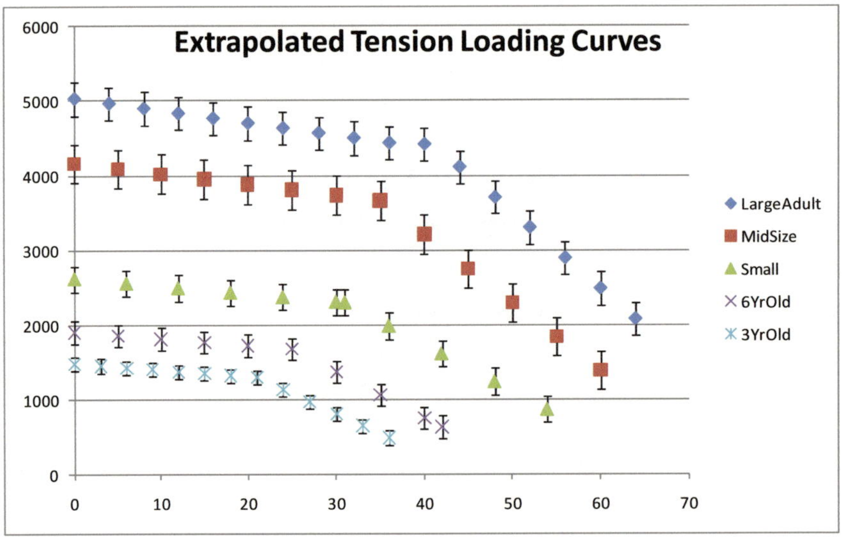

Figure 9. Extrapolated Tension Loading Data

From the standard deviations, ranges of loading were developed for each of the following three scenarios: (i) no injury or minor injury will occur, (ii) A moderate injury is likely to occur and (iii) an unsurvivable injury is likely to occur. For the purposes of comparison, no/minor injury in this analysis refers to an injury that rates from 0 to 1 on the AIS scale.

Similarly, a moderate injury refers to AIS 2-5 and an unsurvivable injury, to AIS 6. These three scenarios were plotted for compression, tension and shear loading for the five different body types previously mentioned.

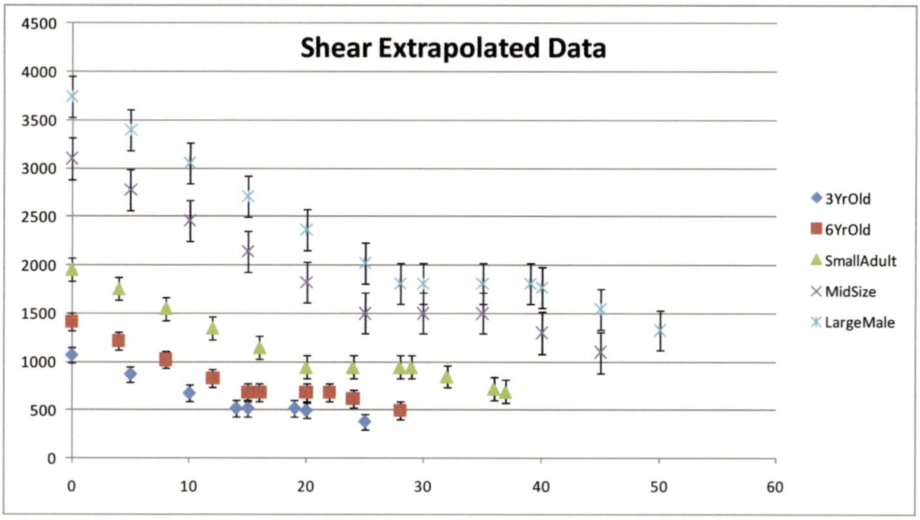

Figure 10. Extrapolated Data for Shear Loading

4. Probability of risk

With the assistance of an Excel Add-In called RiskAmp, numerous Monte Carlo Simulations were set up to study the probability of the five different dummy types being exposed to various compressive, tensile and shear forces. Because of their reliance on repeated computation of random or pseudo-random numbers, these methods tend to be used when it is unfeasible or impossible to compute an exact result with a deterministic algorithm [22]. Applied forces were randomly generated for 1000 simulations. The simulation means and standard deviations were studied. Ranges of force values known to produce responses in the 3 "Injury Zones" were tested against the simulation means to determine the probability of exposure to the varying degrees of compressive, tensile and shear forces.

The risk of injury of being exposed to a force that would place the dummy in each of the three previously discussed injury zones (No Injury Likely, Moderate Injury Likely and Unsurvivable Injury Likely) were developed thereafter.

5. Results

Axial loading, whether tension or compression, can pose a significant risk of injury as seen by the ISO-13232 (Figure 11) testing and analysis procedures [23]. Figure 12 shows the axial neck force time responses as measured in a laboratory head impact test and computer simulation. In figure 12, it can be seen that after only 5 milliseconds, the largest compressive force is exerted on the neck of the rider. After only 15 milliseconds, the rider is then exposed to the highest tensile forces; a direct result of the neck rebounding from compression. And finally, after approximately 30-35 milliseconds, the reactive forces level off. This plot illustrates that the majority of force exposure in impact scenarios occurs within the first 30 milliseconds. It is thus important to focus on the risk of injury during that time frame.

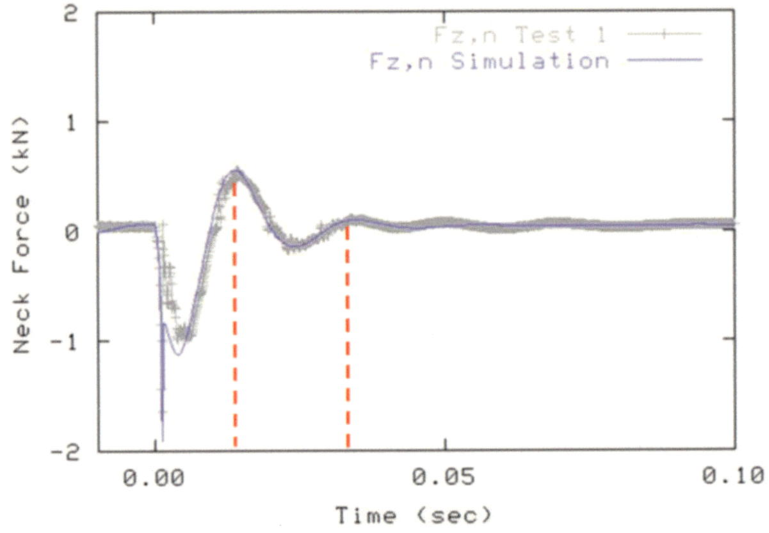

Figure 11. ISO-13232 Axial neck Force Time Responses Measured in a Laboratory Head Impact Test and Computer Simulation [23]

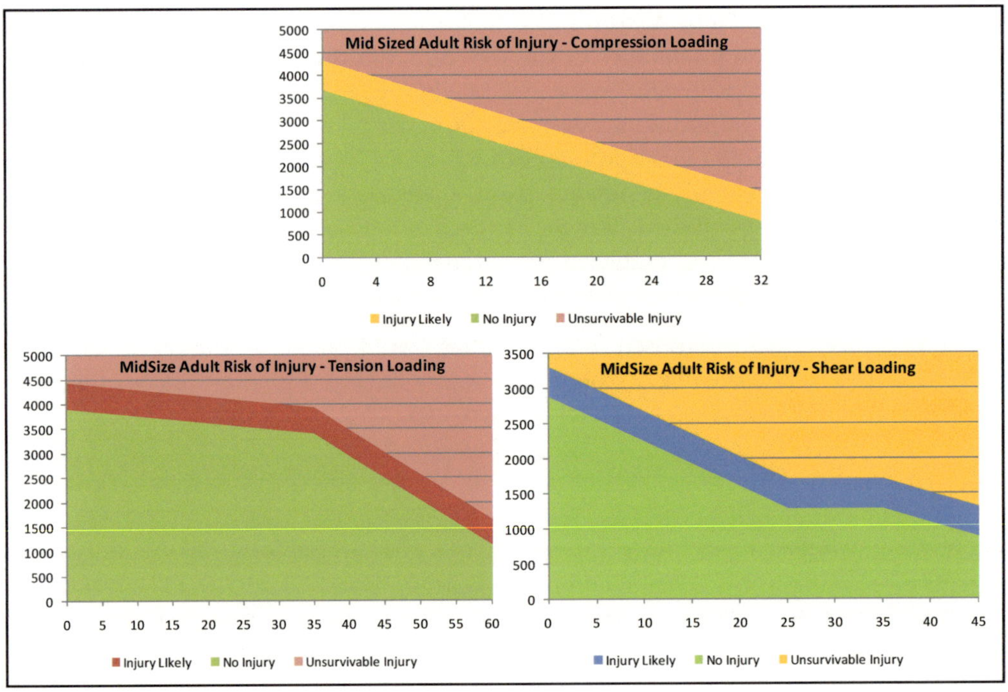

Figure 12. Mid Size Adult Risk of Injury Criteria

Trends are apparent when the five mannequins are all compared under the same type of loading. Under compression, they all exhibit a linear tolerance to loading. For the mid size adult, an instantaneous force between 3700 and 4300 N indicates a significant injury is likely. Anything above 4300 N, illustrates a higher risk of an unsurvivable injury. The linear descending trend remains apparent throughout the first 32 milliseconds of force duration. At 32 milliseconds, it only takes a force between 1000 N and 1600 N to expose the adult to a risk of significant injury. At 1600 N, applied for 32 ms means that an adult is at very high risk of an unsurvivable injury. As the applied compressive load increases with respect to time, the probability of injury linearly increases.

When exposed to tension, a descending, "bi-linear" relationship is seen. Instantaneous forces below 3900 N pose no significant risk of injury. Anything between 3900 N and 4500 N indicates a considerable risk of injury. Finally anything above 4500 N illustrates a dangerous risk of unsurvivable injury. A slightly decreasing linear trend occurs between 0 and 35 ms. At this point in load duration, the linearity changes, the decreasing slope becomes aggressive and the risk of injury becomes more severe. At 35 ms, it takes between 3500 N and 4100 N to see a significant risk of injury, but as time increases to 60 milliseconds, only a force of 1100 N is needed to generate a significant risk of moderate injury. In the first half of the plot, a very gradual decrease in load tolerance with respect to time is evident, but after a load duration of 35 milliseconds, the slope of linearity significantly decreases, indicating that a much higher probability of injury exists.

The shear loading relationship increases in complexity, as it is "tri-linear" in nature. The first third of the plot demonstrates a strong descending linear relationship. An instantaneous force of 2900 N poses a good chance for a moderate to severe injury. Any force above 3300 N instantaneously applied shows a very high risk for an unsurvivable injury. Initially, as with all the mannequin types, a gradual decrease in shear loading tolerance is evident. After approximately 25 milliseconds, the tolerance to shear force plateaus. From 25 to 35 milliseconds of load duration, a force of only 1300 N is needed to substantiate the chance for a moderate injury. Anything above 1800 N indicates that the risk of an unsurvivable injury occurring is quite high. Any load duration beyond 35 ms, sees another decreasing linear trend. It is at the point of 45 ms, that a force of only 900 N is needed to pose significant threat of injury on the neck. This curve behavior illustrates the neck's ability to resist twisting just prior to complete fracture.

Because the probability of injury trends for the five mannequins were similar, only the loading curves for the mid size adult are shown (Figures 12).

Injury due to shear loading seems to happen at much smaller loads for all body types than injuries caused by the other types of loading. For example, the 3 year old experiences instantaneous injury from a tension or a compression load of approximately 1400 N; however the shear load needed to induce injury is only approximately 1000 N. As age and physique increase, a body's tolerance to loading also increases. For example, the 6 year old mannequin does not experience instantaneous injury due to compressive loading until a load of 1700 N is applied. This is higher than the load that a 3 year can forebear, but much less than the 4400 N needed to cause compressive injury to a large adult.

The risk of injury has been presented graphically with respect to the types and magnitudes of forces that are more likely to cause injury, based on age and load duration. The succeeding tables depict the average probability of injury, based on body and loading types. For each Monte Carlo Simulation performed, probability of injury was calculated for a variety of time steps, starting with 0 milliseconds and concluding with a time value known to cause serious injury (i.e. for compressive loads, a concluding time of 32 milliseconds was used for all five mannequin types). The probabilities listed in Tables 3-6 were calculated by averaging the results of the Monte Carlo Simulations for each time step. These values were averaged for the purpose of data consolidation. As can be seen in Tables 4-6, the bigger the mannequin, or human body type, the higher the probability that no injury will occur, and the less likely that an unsurvivable injury will take place.

Hybrid III Family			
Mannequin	Height	Weight	
	(ft, in)	(lbs)	(kg)
3 Yr Old	--	33	15
6 Yr Old	--	47	21
Small Adult	5'0"	110	50
Midsize Adult	5'10"	170	77
Large Adult	6'2"	223	100

Table 3. The Hybrid III Family of Mannequins Data

Probability of Compressive Injury			
Dummy	No Injury	Injury Likely	Unsurvivable
3 Yr Old	27/100	9/100	64/100
6 Yr Old	32/100	11/100	57/100
Small	29/100	12/100	59/100
Midsize	45/100	17/100	38/100
Large	49/100	18/100	32/100

Table 4. Probability of Injury Caused by Compression

Probability of Tensile Injury			
Dummy	No Injury	Injury Likely	Unsurvivable
3 Yr Old	39/100	7/100	54/100
6 Yr Old	41/100	11/100	49/100
Small	61/100	12/100	27/100
Midsize	67/100	11/100	23/100
Large	79/100	7/100	14/100

Table 5. Probability of Injury Caused by Tension

Probability of Shear Injury			
Dummy	No Injury	Injury Likely	Unsurvivable
3 Yr Old	24/100	9/100	67/100
6 Yr Old	30/100	10/100	60/100
Small	35/100	11/100	55/100
Midsize	45/100	15/100	40/100
Large	48/100	14/100	38/100

Table 6. Probability of Injury Caused by Shear Force

6. Discussion

In each of the five cases, for the five different body types, data was compiled from previously published loading curves to determine what type of loading has the chance to cause a significant cervical spinal injury. Shear loading produces a much higher risk of injury on the neck at much lower loads when compared to compressive and tensile loading. If, for example, a compressive load were instantaneously applied to a mid size adult, but wasn't maintained, it takes approximately 3600 N to even enter the region that indicates there is a potential for significant injury. If that force were, however, a shear force, it only takes approximately 2900 N to enter that region (Figure 12). In each of the five body types, the safe and seriously injured regions are well defined. The middle regions, however are those most uninvestigated. They illustrate that injury is fairly likely to occur, but do not illustrate the severity of the injury depending on where one lies within that region. The probability definitions (Tables 4-6) supplement these risk injury curves by providing some scenarios in which various compressive, tensile and shear forces can cause significant injury.

It was important to recognize that not all body types are commonly subjected to forces of dangerous magnitude. For example, a 3 year old would most likely be secluded from situations that could cause him or her to endure such forces, unless the situation was unexpected such as an automobile accident. A large adult, however might perform everyday lifting and pushing tasks that make him or her more likely to encounter high compressive, tensile or shear forces. To take this form of exposure into consideration, Pert distributions (instead of normal distributions) were utilized in the Monte Carlo simulations. Pert distributions help identify associated risk and the likelihood of particular situations occurring, such as various types of people enduring any of a combination of applied loads. This made the results of the simulations more accurate and appropriate for this particular study.

Verifying these Monte Carlo simulations with experimental data was the next step with respect to this research. Cadaveric data was obtained from the literature to validate the accuracy of the Monte Carlo models [9]. As can be seen in Figure 12, the higher the applied force, the greater the probability of injury. Probability of injury was therefore plotted with respect to applied force for all five mannequin types under all three types of loading. A relationship was developed that indicated how probability of injury changed with respect to applied force. To relate the mechanisms of injury to a more common means of risk evaluation, equations were developed relating the applied force to the Abbreviated Injury Scale (AIS). The probability of achieving an injury with an AIS score between 2 and 5 can now be determined simply by knowing the applied force and using the following equations:

$$P(AIS \geq 2) = \left(\frac{1}{1 + e^{(2.056 - 1.1955(N_{ij}))}}\right) \times 100\%$$

$$P(AIS \geq 3) = \left(\frac{1}{1 + e^{(3.227 - 1.969(N_{ij}))}}\right) \times 100\%$$

$$P(AIS \geq 4) = \left(\frac{1}{1 + e^{(2.693 - 1.196(N_{ij}))}}\right) \times 100\%$$

$$P(AIS \geq 5) = \left(\frac{1}{1 + e^{(3.817 - 1.196(N_{ij}))}}\right) \times 100\%$$

In the above equations, N_{ij} refers to the normalized force. For the purposes of risk analysis, the normalized force is identified as the applied force, divided by the critical force, or forced deemed as having the minimum magnitude needed to induce injury.

Both cadaveric and mannequin data of applied force and the resulting probability of various AIS scores was plotted for each type of force. The results of these plots can be seen in Figures 13-15 for compression, tension and shear loading, respectively. In Figure 13, it can be seen that it takes approximately 4000 N of compressive force to generate a 30% risk of an AIS≥2 injury.

Comparatively, from Figure 15, it only takes approximately 3000 N of shear to generate that same 30% risk of an AIS≥2 injury. This again confirms that shear force poses a much higher probability of significant injury over the other types of forces at much lower magnitudes.

This collection of plots sets a solid foundation for understanding what types of loading over what periods of time can injure someone, as well as the probability of someone experiencing different magnitudes of force over different time intervals. It is evident in the comparison of mannequin to cadaver data that mannequin testing does supply an accurate representation of a body's response to the various types of loading. Additionally it provides a more realistic means for studying a person's tolerance to force. The graphic schemes defined in this research have helped to identify the most common mechanisms of cervical spine injury.

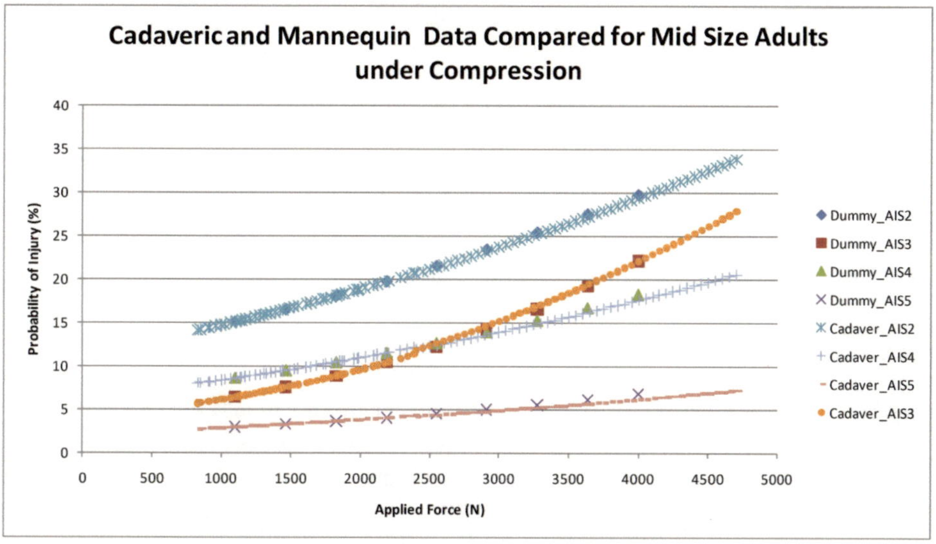

Figure 13. Figure 13: Cadaver and Mannequin Compression Loading Data

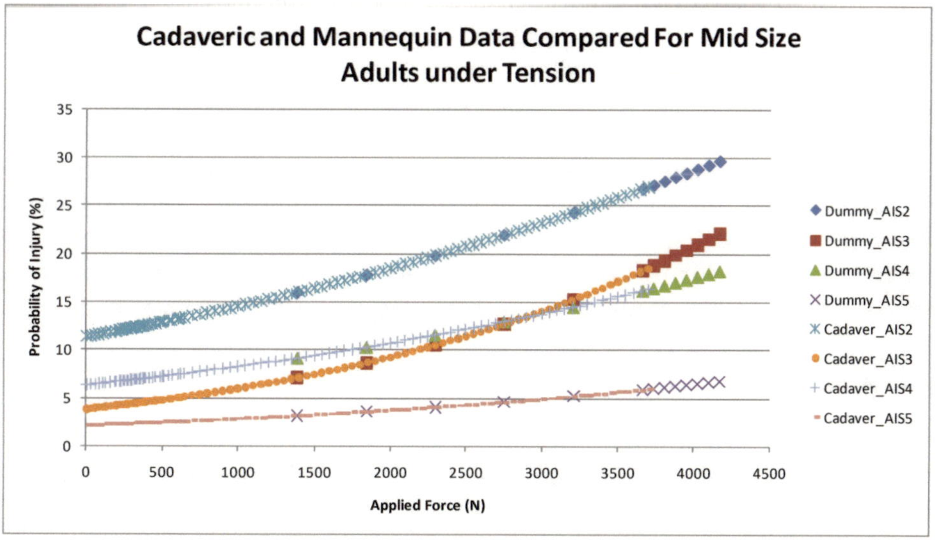

Figure 14. Figure 14: Cadaver and Mannequin Data for Tension Loading

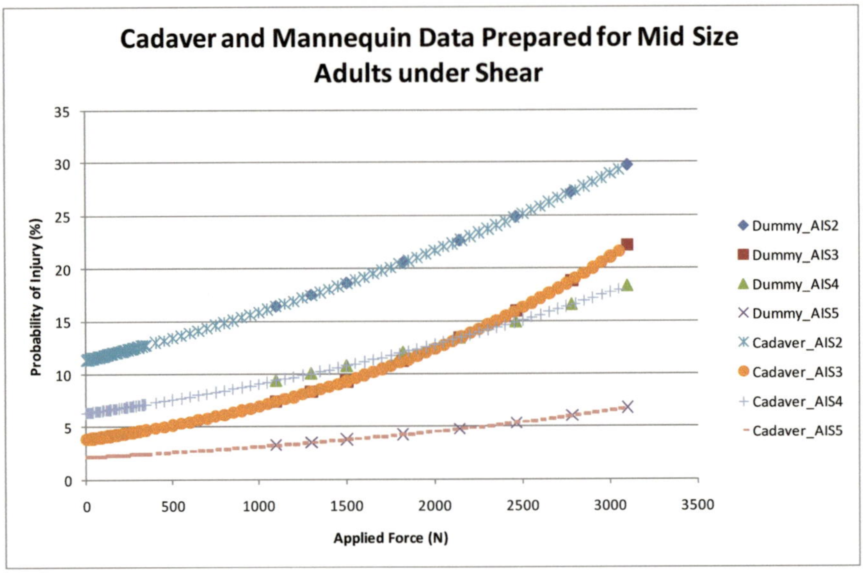

Figure 15. Cadaver and Mannequin Data for Shear Loading

7. Conclusions

This research presents a summary of various conditions under which the cervical injury takes place. The automobile accidents proved the most frequent cause of cervical injury in both Canada and the United States. Data from these countries was analyzed and envelopes where injury takes place graphically presented. Five different body types were used to help illustrate what types of loading are most likely to cause injury when they are applied over varying amounts of time. In all cases, shear loading causes the most risk for injury at smaller loads when compared to tensile and compressive loads. Although injury curves establish a good starting point for identifying risks, more research is needed to fully understand to what degree someone can be injured when exposed to these types of loads. Monte Carlo Method was used to simulate the probability of someone being exposed to different magnitudes of forces and different types of loading and their representative AIS scores in terms of new mathematical equations. These models were then developed to relate applied force to probability of injury at differing levels on the AIS. For all body types, the shear force mechanism posed the highest probability of injury, needing far less force to generate the potential for serious damage. It was also determined that the comparison of mannequin and cadaveric data, that mannequin testing provides an accurate means for assessing a person's probability of injury for all body types.

Author details

Mary E. Blackmore, Tarun Goswami* and Carol Chancey
Spine Research Group, Biomedical, Industrial and Human Factors Engineering Department, Wright State University, U.S.A.

* Corresponding Author

8. References

[1] BrainandSpinalCord.org. Resources and Information for Brain and Spinal Cord Injury Survivors. Retrieved on December 9th, 2010

[2] Cheng R., Y. K. (1982). Injuries to the Cervical Spine Caused by a Distributed Frontal Load to the Chest. Proceedings of the 26th Stapp Car Crash Conference SAE Paper 821155 , 899-938.

[3] Clemens HJ., B. K. (1972). Experimental Investigation on Injury Mechanisms of Cervical Spine at Frontal and Rear Front Vehicle Impacts. SAE Paper 720960 , 78-104.

[4] Doherty, B. E. (1993). A Biomechanical Study of Odontoid Fractures and Fracture Fixation. Spine , 18 (2), 178-184.

[5] Fielding JW., C. G. (1974). Tears of the Transverse Ligament of the Atlas. Journal of Bone and Joint Surgery , 56A (8), 1683-1691.

[6] Goel VK., W. J. (1990). Ligamentous Laxity Across the C0-C1-C2 Complex: Axial Torque-rotation Characteristics Until Failure. Spine , 15, 990-996.

[7] Goldberg W., M. C. (2001). Distribution and Patterns of Blunt Traumatic Cervical Spine Injury. Annals of Emergency Medicine , 38 (1), 17-21.

[8] Lars, B. K. (2006). Acute Spinal Injuries: Assessment and Management. Emergency Medicine Practice , 8 (5), 1-28.

[9] Liu, Y. King, Krieger, K.W., Njus, G., Ueno, K., Connors, M.P., Wakano, K., Thies, D. Cervical Spine Stiffness and Geometry of the Young Human Male. Air Force Aerospace Medical Laboratory. Tulane University School of Medicine, New Orleans, LA. Approved for Public Release January 19, 1983.

[10] Maiman DJ., S. A. (1983). Compression Injuries of the Cervical Spine: A Biomechanical Analysis. Neurosurgery , 13 (3), 254-260.

[11] Mertz HJ., P. L. (1971). Strength and Response of the Human Neck. Proceedings of the 15th Stapp Car Crash Conference , 207-255.

[12] Myers BS., M. J. (1991). The Influence of End Condition on Human Cervical Spine Injury Mechanisms. Proceedings of the 35th Stapp Car Crash Conference , 391-400.

[13] Myers BS., M. J. (1991). The Role of Torsion in Cervical Spinal Injuries. Spine , 16 (8), 870-874.

[14] Nahum Alan M., M. J. (2002). Accidental Injury: Biomechanics and Prevention. New York: Springer.

[15] Pintar FA., Y. N. (1995). Cervical Vertebral Strain Measurements under Axial and Eccentric Loading. Journal of Biomechanical Engineering , 474-478.

[16] Portnoy HD., M. J. (1972). Mechanism of Cervical Spine Injury In Auto Accident. Proceedings of the 15th Annual Conference of the American Association for Automotive Medicine , 58-83.

[17] Reid DC., S. L. (1989). Spine Fractures in Winter Sports. Sports Medicine , 7 (6), 393-399.

[18] Sekhon, L. F. (2001, December 15). Epidemiology, Demographics and Pathophysiology of Acute Spinal Cord Injury. Spine , S2-12.

[19] Shea M., W. R. (1992). In Vitro Hyperextension Injuries in Human Cadaveric Cervical Spines. Journal of Orthopedic Research , 10 (6), 911-916.

[20] Shield LK., F. B. (1978). Cervical Cord Injury in Sports. Physical Sports Medicine , 6, 321-326.

[21] Spinal Cord Injury: Facts and Figures at a Glance. (1999, April). Retrieved March 2010, from National Spinal Cord Injury Statistical Center: www.spinalcord.uab.edu

[22] Todinov, M. (2005). Reliability and Risk Models: Setting Reliability Requirements. Chichester, West Sussex, England: John Wiley & Sons Ltd.

[23] Van Auken, Michael R., Zellner, John W. and Terry Smith. Development of an Improved Neck Injury Assessment Criteria for the ISO 13232 Motorcyclist Anthropometric Test Dummy. International Motorcycle Manufacturers Association, Switzerland. (2005) Paper No. 05-0227

[24] Wismans JS., S. C. (1983). Performance Requirements of Mechanical Necks in Laterial Flexion. Proceedings of the 27th Stapp Car Crash Conference 137.

[25] Yoganandan N., H. M. (1989). Epidemiology and Injury Biomechanics of Motor Vehicle Related Trauma to the Human Spine. Proceedings of the 33nd Stapp Car Crash Conference , 1102-1110.

Section 3

Gait Behavior

Chapter 8

Methodology for the Assessment of Joint Efforts During Sit to Stand Movement

Maxime Raison, Maria Laitenberger, Aurelie Sarcher,
Christine Detrembleur, Jean-Claude Samin and Paul Fisette

Additional information is available at the end of the chapter

http://dx.doi.org/10.5772/49996

1. Introduction

The sit to stand (STS) analysis and particularly the 5-repetition sit-to-stand test (FRSTST) introduced by Bohannon [1] are widely used measurements of functional strength and disability level of young and elderly subjects. For example in rehabilitation and orthopedics, these tests are mainly used for the functional evaluation of :

- children with cerebral palsy, for which the FRSTST was found a reliable and valid test to measure functional muscle strength in children with spastic diplegia in clinics [2, 3];
- older adults, for which the FRSTST test-retest reliability can be interpreted as good to high in most populations and settings [1];
- subjects with Parkinson's disease [4];
- paraplegic subjects [5];
- subjects with multiple sclerosis [6];
- above knee amputees [7] and unilateral transtibial amputees [8];
- subjects with rheumatoid arthritis [9] or alterations in advanced knee osteoarthritis [10];
- post-stroke subjects [11].

The STS analysis also currently helps to develop :

- STS assistive device for the elderly and disabled [12];
- STS and gait support system for elderly and disabled [13], and also handrail positions and shapes that best facilitate STS movement [14];
- car cockpits taking into account the comfort analyzes of subject seated in a car [15, 16] or on a simple seat [17, 18].

Usually, the functional strength and the disability level during STS are evaluated by calculating the total forces of hip and knee extensors [1] and the center of mass (COM) accelerations [19]. Nevertheless, it is known that determining with accuracy the kinematics (including the COM) and dynamics (including the joint forces and torques) in the human body is still a great challenge in biomechanical modeling [20]. Consequently, the aim of the present study consists in presenting a rigorous methodology for the non-invasive assessment of joint efforts and the associated kinematic variables during STS movement. This method is based on a three-dimensional dynamical inverse model of the human body. Like other classical *dynamical inverse* analyzes [21–25] in biomechanics of motion, the model proposed here [18] uses measurements of external interactions (forces $\mathbf{F}_{ext}$ and torques $\mathbf{M}_{ext}$) between the body and its environment, and also measurements of the system configuration x_{exp}. The corresponding joint coordinates q are numerically determined by a kinematic identification process, and the corresponding velocities $\dot{q}$ and accelerations $\ddot{q}$ are presently estimated from the q, using a numerical derivative. Finally, the model provides the joint interactions with the use of a symbolically generated recursive Newton-Euler formalism [26, 27].

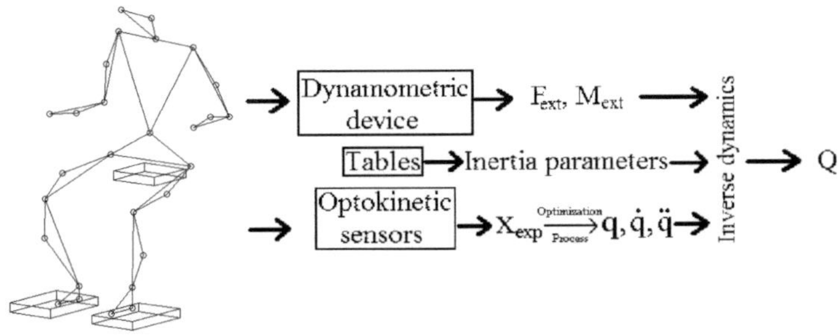

Figure 1. Principle of the inverse dynamical model: from the experiment to the vector Q of the joint efforts.

This model is applied to experiments of STS : the subject, initially seated, is asked to get up without moving the feet, and without arm or hand contact with the environment or with any part of the body. In this paper, both postural behaviors of slow and fast STS are analyzed and compared.

2. Material and methods

First, this section summarizes the features of the proposed human body model and describes the corresponding experimental set-up and process. Second, a preliminary calculation defines the centers of mass and centers of pressure of the model, and also develops the relation between their local and global components: these variables are known as diagnostic tools in rehabilitation and physical ergonomics [28–30], and useful for the present model analysis. Third, the theoretical investigation will develop both kinematic and dynamical analyzes related to this model, and both analyzes will be applied to the STS.

2.1. Model features and hypotheses

The proposed human body model is composed of 28 position sensors (Fig. 2), defining 13 rigid bodies: the head, both upper arms, both lower arms, the trunk, the pelvis, both thighs, both shanks, and both feet. Each of the 13 bodies is defined by three position sensors, in order to know the three-dimensional configuration of each body. Further, these bodies are linked by spherical joints corresponding to 12 anatomical landmarks (referring to [31]): the C7 vertebra, both shoulders (acromioclavicular joints), both elbow joint centers, the sacrum, both greater trochanters, both knee joint centers, both lateral heads of the malleolus. Consequently, the system is fully described by a total of *13 (bodies)* $\times$ *6 variables - 12* $\times$ *3 spherical joint constraints = 42 generalized coordinates*, representing the 42 degrees of freedom of the model. As shown in Fig. 1, the inverse dynamical model provides the column vector Q of joint forces and torques, using three sets of inputs:

1. The external forces and torques.
2. The inertia parameters.
3. The joint coordinates, velocities and accelerations.

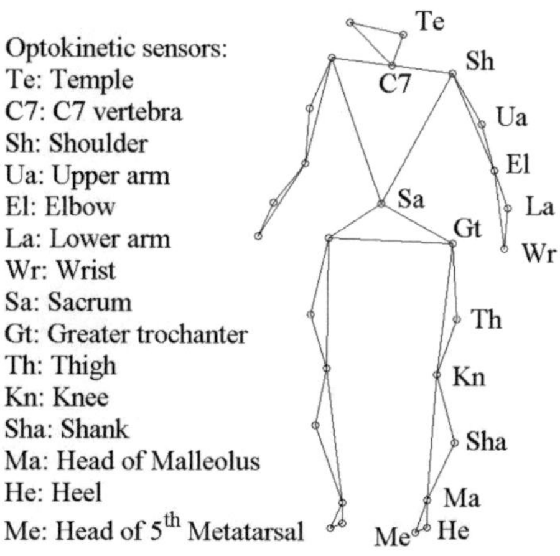

Figure 2. Human body model, featuring the 28 optokinetic sensors that define the 13 articulated rigid bodies, each defined by three points, articulated via 12 spherical joints.

A few characteristics and assumptions must be formulated about these three sets of inputs:

- The external forces $\mathbf{F}_{ext}$ and torques $\mathbf{M}_{ext}$ between the body and its environment are measured by a dynamometric device. The external pure torques are not considered.
- The body inertia parameters, i.e. the masses m_i, moments of inertia I_i and center of mass positions $\overrightarrow{OM_i}$ of the ith body member (i = 1,...,13) are taken from the inertia tables of de Leva [31] (1996) readjusted from the Zatsiorsky-Seluyanov's mass inertia parameters [32] (1990). The inertia parameter identification is not part of this research: indeed, previous investigations [33] showed that non-invasive in-vivo identifications of the body parameters are presently inappropriate to the human body dynamics, because the resulting body parameters have significant errors due to experimental errors in the input data, such as the body configuration, or the external force and torque measurements.
- The system configuration, i.e. the experimental absolute coordinates x_{exp} of the reference points, are measured by the 28 optokinetic sensors. The corresponding joint coordinates q are numerically determined by a kinematic identification process and the corresponding velocities $\dot{q}$ and accelerations $\ddot{q}$ are presently estimated from the q by a numerical derivative

using finite differences. Considering the joint kinematics, we are aware that more adequate joint models could be used: in particular, previous studies [34, 35] have developed more complex three-dimensional joints for the knee and the shoulder. The present model has been implemented with spherical joints but will be extended to include more involved joints in the future. Further, the results of the kinematic analysis for this experiment show that the spherical joints considered here sufficiently fit the considered motion (see Section 3.1).

2.2. Experimental set-up and procedure

Let us consider the system reference frame $[\hat{I}]$, located at a fixed point O on the laboratory floor (Fig. 3). In this reference frame, the motion measurement set-up consists of optokinetic sensors and six infra-red cameras ($Elite - BTS^{TM}$), that estimate the coordinate vectors $\overrightarrow{OX}_{exp,n} = [\hat{I}]^\top x_{exp,n})$ of the joint reference points, i.e. of the optokinetic sensors. Further, the interaction measurement set-up consists of two force platforms at the feet contact and one force platform at the seat, for the determination of the horizontal and vertical interaction forces $\mathbf{F}_{ext} = [\hat{I}]^\top F_{ext}$ and torques $\mathbf{M}_{ext} = [\hat{I}]^\top M_{ext}$ between the body and these platforms. The three independent platforms are composed of four force sensors [36], designed by our laboratory, and located at the edges of these platforms. The device provides a total number of *3 platforms* $\times$ *4 force sensors* $\times$ *3 force components = 36 force components*. All data are sampled at 100 Hz, using an adaptive low-pass numerical filter (implemented by $Elite - BTS^{TM}$).

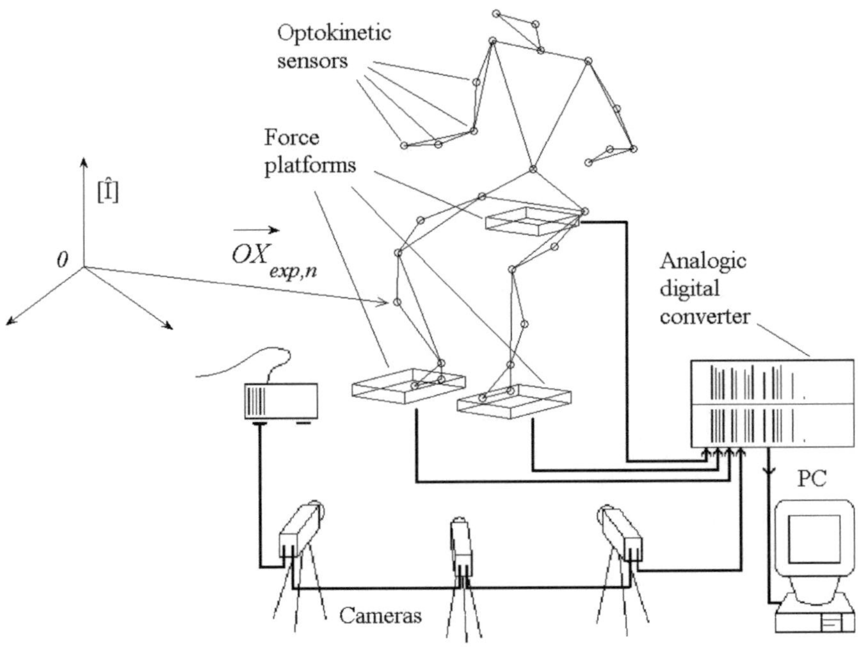

Figure 3. Experimental set-up, related to the system reference frame $[\hat{I}]$, located at a fixed point O on the laboratory floor. $\overrightarrow{OX}_{exp,n}$ represents the coordinate vectors of the optokinetic sensors.

The experiments were performed by one person related to our laboratory, who gave his informed consent to perform the experiments. Note that further experiments of STS are presently performed in order to discuss the repeatability of the data and results for several subjects and several behaviors of STS.

At the beginning of each test, the subject is seated as shown in Fig. 3. Then the subject is asked to get up from the seat. During the whole experiment, the observers check that:

- the subject do not move feet, in order to obtain a good repeatability of the initial and final body configurations;
- the subject has neither arm nor hand contact with the environment or the rest of the body.

Two behaviors of STS are analyzed and compared in this paper, in order to compare a "slow" and a "fast" STS. For both tests, the time evolution of the body motion permits the definition of three phases:

1. The initial phase: the subject is seated, it is assumed that the subject is at an equilibrium state, i.e. the subject is only performing forces necessary to maintain his initial posture.
2. The transient phase, composed of two sub-phases: a first transient sub-phase when the subject begins to get up and the subject thighs are still in contact with the seat; a second transient sub-phase, when the subject continues to get up without seat contact.
3. The final phase: the subject maintains his standing-up position; it is assumed that this is the second equilibrium state of the subject during the test.

2.3. Center of mass and center of pressure

This section defines the centers of mass and centers of pressure of the proposed model, and also develops the relation between their local and global components [29, 30].

2.3.1. Centers of mass

The position of the *global center of mass* (GCOM) of the human body can be written as follows:

$$\overrightarrow{OM} = \frac{\sum_{i=1}^{13} m_i \overrightarrow{OM_i}}{\sum_{i=1}^{13} m_i} \tag{1}$$

where

- $\overrightarrow{OM_i}$ is the position vector of the *local center of mass* LCOM of the ith body member (i=1,. . . ,13); the values of $\overrightarrow{OM_i}$ are estimated from the human body configuration and the inertia tables of de Leva [31];
- m_i is the mass of the ith member (i=1,. . . ,13); the values of m_i are estimated from the inertia tables.

Remember that the integration of the platform force data provides more accurate values of the GCOM variations [37], which are used as diagnostic tools in rehabilitation and physical ergonomics. However, the GCOM calculated by this method is equal to the actual GCOM plus one undetermined constant value. Further, it was shown for instance that the differences between the GCOM estimated by these two methods are less than 0.3% height in all 3 components for able bodied subjects [37]. Consequently, the upper definition of the GCOM is preferred to estimate the actual GCOM value of the present human body model.

2.3.2. *Centers of pressure*

For each force platform, the *local center of pressure* (LCOP) components, related to the system referential point O, can be determined from the platform force data, using the following definition:

$$\overrightarrow{OP}_j = (X_{P_j}, Y_{P_j}, Z_{P_j}) = (-\frac{M_{Y_j}}{R_{P_{j,z}}}, \frac{M_{X_j}}{R_{P_{j,z}}}, H_j) \quad (2)$$

where

- the index j indicates the platform: j = 1, 2 or 3 for the left foot platform, the right foot platform or the seat platform, respectively;
- $R_{P_{j,z}}$ is the vertical component of the force on the jth platform;
- M_{X_j} and M_{Y_j} are anterior-posterior and lateral components, respectively, of the resulting moment on the jth platform, related to the reference O;
- H_j is the measured height of the jth platform; H_j is the assumed to be constant during the experiment.

The *global center of pressure* (GCOP) [29, 30] is defined as the weighted sum of the LCOP on every contact platform. Its expression related to the system referential point O is given by :

$$\overrightarrow{OP} = \frac{\sum_{j=1}^{3} R_{P_j} \cdot \overrightarrow{OP}_j}{\sum_{j=1}^{3} R_{P_j}} \quad (3)$$

where, for the platforms from j = 1 to 3 (i.e. j = 1 for the left foot platform, j = 2 for the right foot platform and j = 3 for the seat platform) :

- the index j indicates the platform;
- $\overrightarrow{OP}_j$ is the vector of position of the LCOP on the jth platform;
- R_{P_j} is the global force data on the jth platform.

Let us note that $\overrightarrow{OP}_i$ and R_{P_j} are totally estimated from the platform force data. In particular, during the second part of the transient phase and the final phase, when there is no contact between the subject thighs and the seat, $R_{P_3} = 0$ and $\overrightarrow{OP}$ does not take into consideration OP_3, which is undetermined from Equation (2).

Finally, both centers of mass and centers of pressure will be presented in the 'Results' Section, because these are useful in rehabilitation and physical ergonomics. However, only the LCOMs, estimated from the system configuration and the tables of inertia, are essential for the implementation of the musculoskeletal analysis presented in Fig. 1.

2.4. Theoretical investigation

The theoretical investigation of the model is developed in two steps :
- First, the joint coordinates q are numerically determined by an identification process that estimates the joint coordinates of the multibody model that best fit the experimental joint positions $x_{exp,n}$.
- Second, the dynamical model provides the vector Q of joint torques during the experiments, using a symbolic generated recursive Newton-Euler formalism. Let us note that this vectorial formulation allows the results to be independent of the angle variable in the spherical joints, whose choice and sequence are rather defined for a methodical implementation than for physiological reasons.

2.4.1. *Kinematic analysis*

The joint coordinates q are numerically determined by an identification process that estimates the joint coordinates of the multibody model that best fit the experimental joint positions $x_{exp,n}$. As proposed by Ref. [20], the optimization problem can be formulated as a nonlinear least-square problem applied for each body configuration, at each time instant t_k, $k = 1, \ldots, T$, where T is the last time sample of each test. Consequently, the cost function $f_{cost}(t_k)$ can be written at each time instant t_k as follows:

$$f_{cost}(t_k) = \sum_{n=1}^{28} |x_{mod,n}(q(t_k)) - x_{exp,n}(t_k)|^2 \quad (4)$$

where

- the index $n = 1, \ldots, 28$ indicates the optokinetic sensor;
- $q(t_k)$ is the joint coordinate vector at the time instant t_k, and is the variable of the optimization process;
- $x_{mod,n}(q(t_k))$ is the cartesian coordinate of the nth optokinetic sensor at the time instant t_k, obtained from the $q(t_k)$, using the forward kinematic model;
- $x_{exp,n}(t_k)$ is the cartesian coordinate of the nth optokinetic sensor at the time instant t_k, provided by the experimental set-up.

Fig. 4 schematically outlines the identification process, which involves two consecutive steps:

1. A pre-process calculates the mean distances l_i between the joints for each of the ith body member, using the experimental joint cartesian coordinates $x_{exp,n}(t_k)$. The reason is that the approach is based on a multibody model, composed of rigid bodies, for which a variable size of the bodies would be irrelevant.
2. The model joint cartesian coordinates $x_{mod,n}$ are given by a forward kinematic model using the l_i distances and an initial value (set to zero) of the joint coordinates $q(t_k)$ that we want to determine. The cost function of this least-square optimization is defined as the sum of the square components of the absolute error vector between $x_{exp,n}(t_k)$ and $x_{mod,n}(q(t_k))$ of the n optokinetic sensors at the time instant t_k. In order to improve the process convergence, the optimal value of $x_{mod,n}(q(t_k))$ becomes the initial condition of the next iteration at the time instant t_{k+1}.

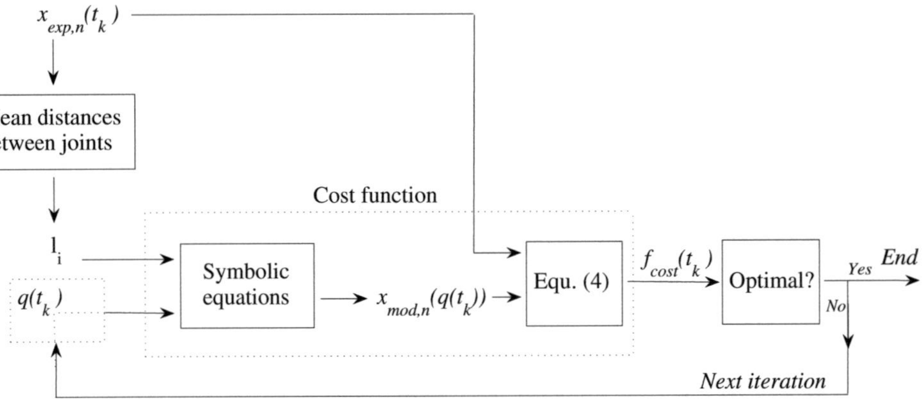

Figure 4. Identification process for a body configuration, at a time instant t_k, as proposed by Ref. [20].

The corresponding velocities $\dot{q}$ and accelerations $\ddot{q}$ are presently derived from the joint coordinates $q(t_k)$ and approximated by finite differences. The noise in $q(t_k)$ could be a significant source of error in the $\dot{q}$ and $\ddot{q}$ estimations, and thus in the dynamical analysis. Consequently, an optimization of $\dot{q}$ and $\ddot{q}$ will probably be suggested in the future. Nevertheless, the fact that the $x_{exp,n}$ are measured using an adaptive low-pass numerical filter and that the $q(t_k)$ are obtained using a kinematic optimization largely improves the $\dot{q}$ and $\ddot{q}$ accuracy.

2.4.2. *Dynamical analysis*

As proposed by Ref. [39], the system dynamical equations are obtained from a Newton-Euler formalism [26, 27]: this algorithm provides the vector Q of internal interaction torques and forces at the joints for any configuration of the multibody system, in the form of an inverse dynamical model (Equation 5), a semi-direct dynamical model (Equation 6) :

$$Q = f(q, \dot{q}, \ddot{q}, F_{ext}, M_{ext}, g) \quad (5)$$
$$= M(q)\ddot{q} + G(q, \dot{q}, F_{ext}, M_{ext}, g) \quad (6)$$

where

- q (42×1) is the vector of the human body joint coordinates, i.e. successively the three angular coordinates for each of the 13 members (*3 (translations for the first member LCOM position) + 13 (members) × 3 (angular coordinates) = 42 components*); the three angular coordinates per member represent the spherical joint; let us note that three translations per joint have been introduced and locked in order to permit the joint force calculations without interfering with the model [26, 27];
- $\dot{q}$ and $\ddot{q}$ (42×1) are the joint velocities and accelerations, respectively;
- F_{ext} and M_{ext} (42×1) are the three-dimensional components of the global external forces and torques applied to each of the body members;
- g (1×3) is the gravity;

- $M(q)$ (42×42) is the positive-definite symmetric mass matrix;
- $G(q, \dot{q}, F_{ext}, M_{ext}, g)$ (42×1) is the dynamical vector containing the gyroscopic, centrifuged and three-dimensional terms resulting from the system configurations, velocities, and also the external forces and torques and gravity applied to the system.

3. Results

In this section, the model is applied to two behaviors of STS, as follows :

- At each time instant t_k, the kinematic optimization problem provides the human body joint coordinates $q(t_k)$ that best fit the experimental joint positions $x_{exp,n}(t_k)$. From these results, a error analysis of the fitted model and a short joint kinematic analysis are developed for two behaviors of STS, defined as a *slow* and a *fast* motion, respectively.
- The inverse dynamical model provides the vector Q of the joint forces and torques for the slow and fast motions, respectively.

Furthermore, segment animations have been developed in order to present the kinematics and dynamics results on the model in a convenient manner. These animations are available on Ref. [40], and a few samples are described in this section.

3.1. Kinematic analysis

In terms of CPU time performance, the kinematic identification process, using $MATLAB^{TM}$ on a Pentium IV 530, 3 GHz processor, requires ca. 30 CPU seconds per 100 experimental samples, i.e. per second of studied motion. Further, the data reconstruction for the animation requires ca. 25 CPU seconds per second of studied motion. Consequently, the total optimization and display process requires ca. 55 CPU seconds per second of studied motion, i.e. in practice, this approximately requires 11 minutes for 10 seconds of motion data recording. Finally, let us note that the identification process time was reduced by 60% using a *mexfunction* from $MATLAB^{TM}$ to C++.

At each time instant t_k, the model joint cartesian coordinates $x_{mod,n}(q(t_k))$ of one behavior (here, the fast motion) can be recalculated in order to build the fitted model (blue in Fig. 5). This fitted model, using purely rigid bodies, can be compared to the purely experimental model (red in Fig. 5), based on the experimental joint cartesian coordinates $x_{exp,n}(t_k)$.

Further, an error analysis provides the global relative errors between $x_{mod,n}(q(t_k))$ and $x_{exp,n}(t_k)$ for the two behaviors of STS, in percentage of the corresponding $x_{exp,n}(t_k)$ at each time instant t_k (Fig. 6). For the fast motion (resp. the low motion), the maximal value of the global relative error is equal to 11.46% (resp. 8.27%) of the corresponding $x_{exp,n}(t_k)$, and the mean value of the global relative error is equal to 0.31% (resp. 0.33%), corresponding to a mean absolute error equivalent to 3.8mm (resp. 3.9mm) in each direction at each joint. In both cases, the error peaks occur during the transient phase of the motion.

Finally, selected results of joint kinematics are presented as follows :

1. The GCOM trajectories are presented (Fig. 7) during the slow a fast motions.

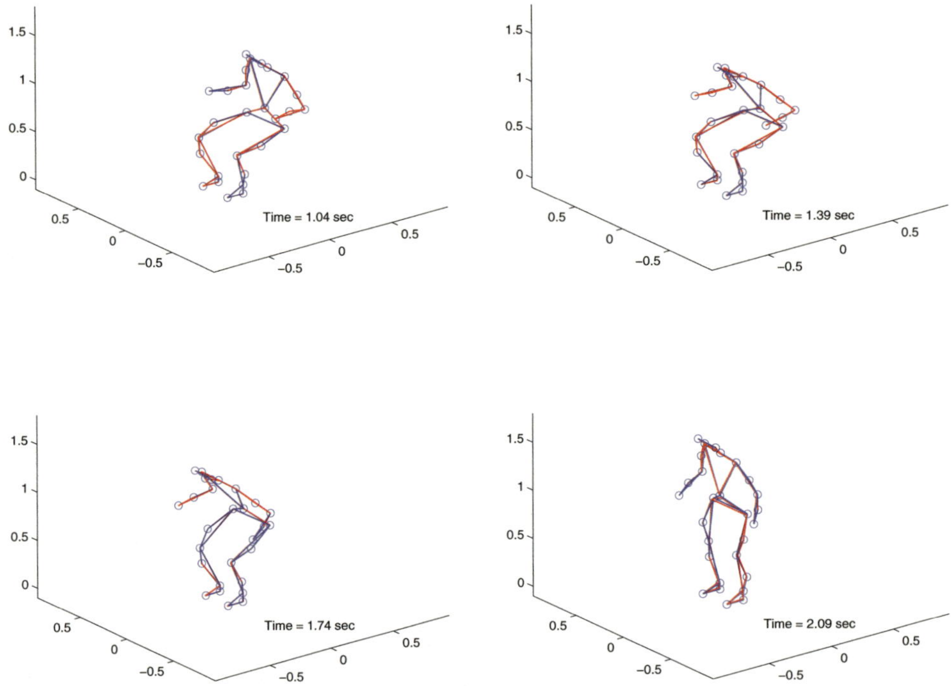

Figure 5. Fast STS: kinematic fit for four time instant t_k; superimposition of the fitted model coordinates $x_{exp,n}(t_k)$ (in blue) and the experimental model coordinates $x_{mod,n}(q(t_k))$ (in red).

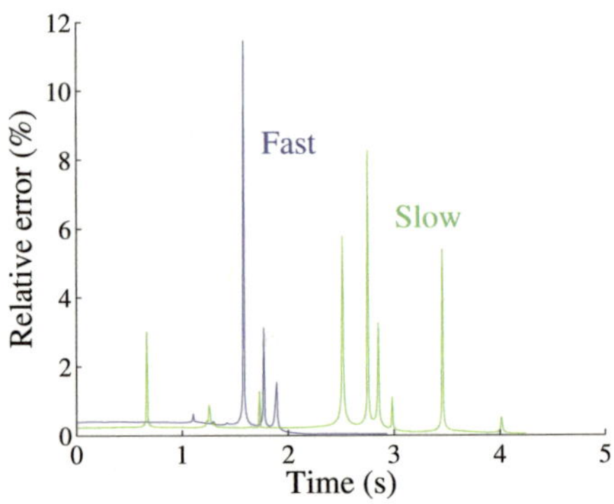

Figure 6. Error analysis: time evolution of the global relative error between $x_{mod,n}(q(t_k))$ and $x_{exp,n}(t_k)$, during the slow (green) and fast (blue) getting-up motions.

2. As an example, the model joint kinematics are compared during the slow a fast motions, for two joints : the consecutive angular coordinates R_3, R_1 and R_2 are described at the

sacrum (Fig. 8), i.e. from the pelvis member to the trunk member, and also at the right elbow (Fig. 9), i.e. from the upper arm to the lower arm.

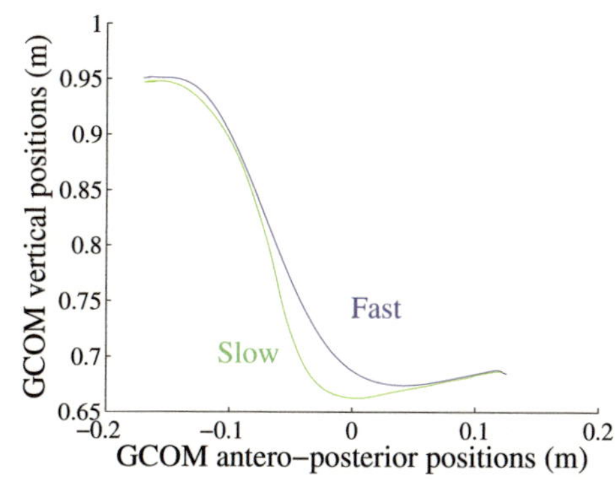

Figure 7. Trajectory of the GCOM during the slow (green) and fast (blue) getting-up motions.

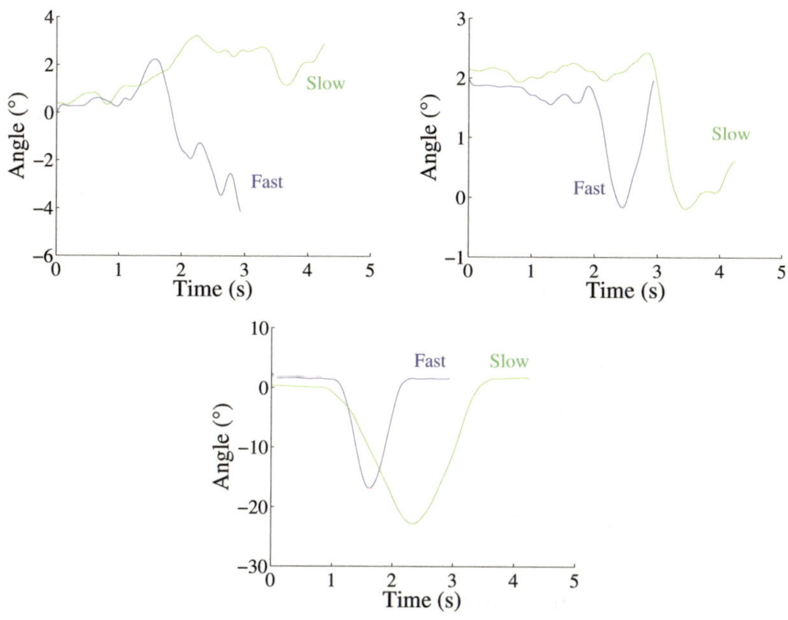

Figure 8. Time evolution of the three consecutive angular coordinates R_3, R_1 and R_2, respectively, at the sacrum: comparison of the slow (green) and fast (blue) getting-up motions.

3.2. Dynamical analysis

On the basis of the reference frame defined in Fig. 3, the dynamical analysis provides the time evolution of the global joint torques (Fig. 10) and forces (Fig. 11), for the slow and fast behaviors.

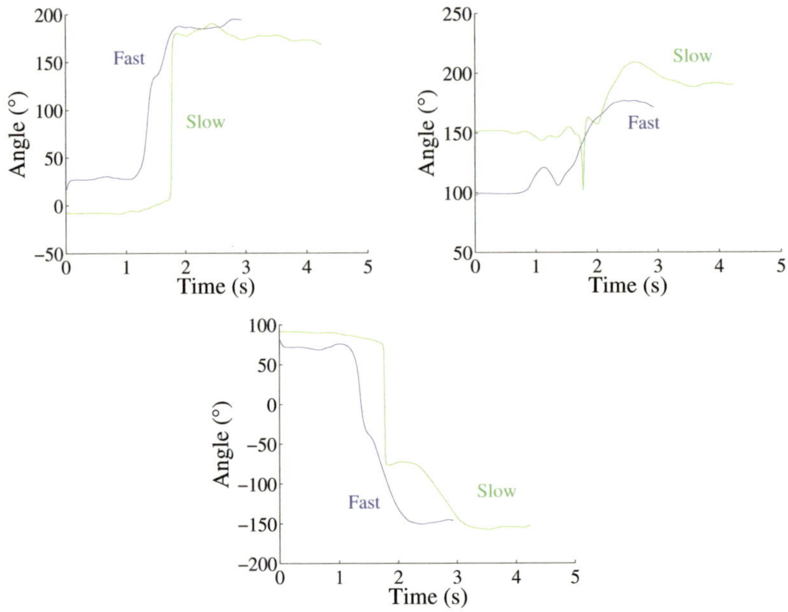

Figure 9. Time evolution of the three consecutive angular coordinates R_3, R_1 and R_2, respectively, at the elbow: comparison of the slow (green) and fast (blue) getting-up motions.

Furthermore, body segment animations have been developed in order to show the evolution of the joint positions, the corresponding global joint torques, and also the local and global centers of mass and pressure. Samples of this animation are presented in Fig. 12, at four time instants t_k during the fast STS behavior.

4. Discussion and conclusion

This section presents the benefits and limitations of this methodology, and also the perspectives for future studies.

4.1. Benefits and limitations

The present inverse dynamical model of the human body coupled with a kinematic identification of the model configurations (Fig. 1) is proposed as an accurate method to estimate the joint efforts in dynamical contexts, as presented from Fig. 1. Nevertheless, three main limitations of the present inverse dynamical model must be discussed.

1. *The geometrical limitation, due to the use of spherical joints :* The results of the kinematic analysis for this experiment show that the spherical joints considered here sufficiently fit the considered motion, with $x_{mod,n}(q(t_k))$ errors corresponding to a mean absolute error inferior to 3.9mm in each direction at each joint. However, using previous investigation results, the present model will be extended to include more involved joints in the future, particularly for the knees [34] and the shoulders [35].

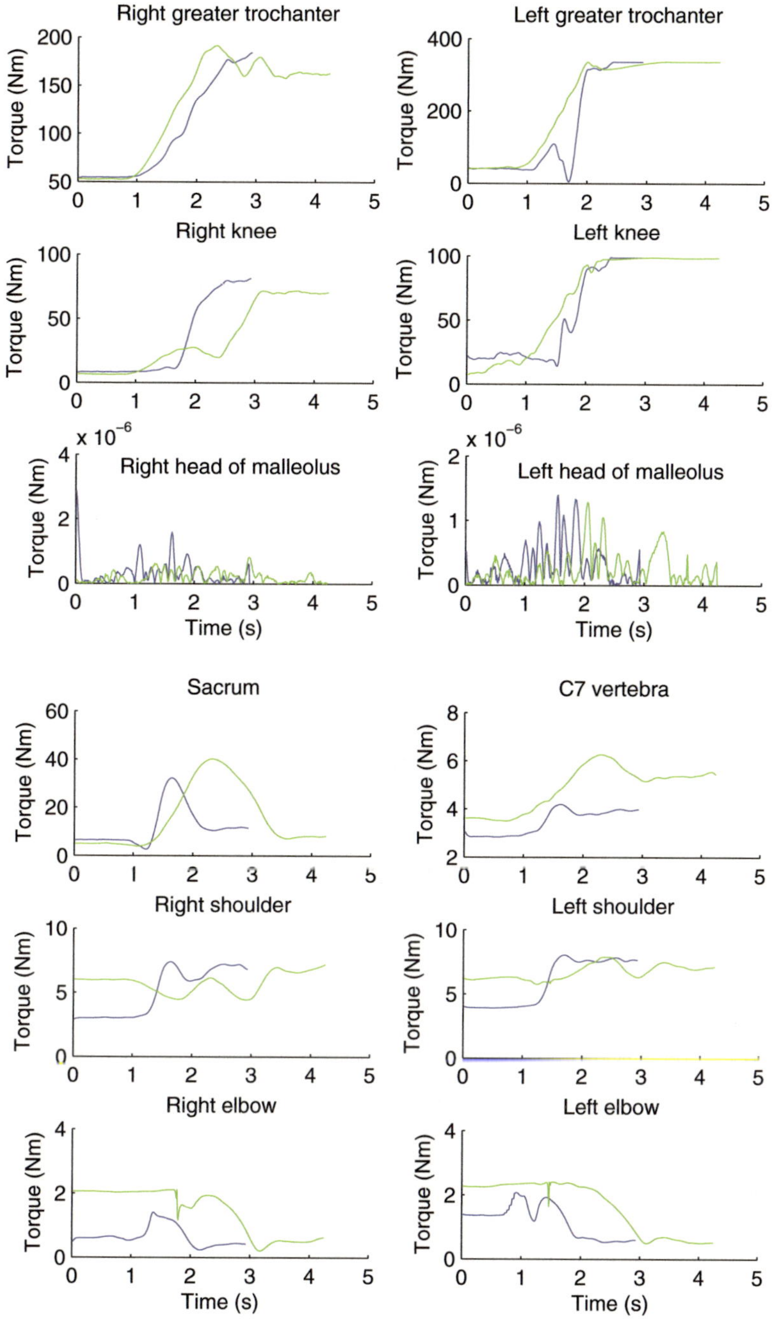

Figure 10. Time evolution of the global joint torques: superposition of the global joint torques at each joint, during the slow (green) and fast (blue) getting-up motions.

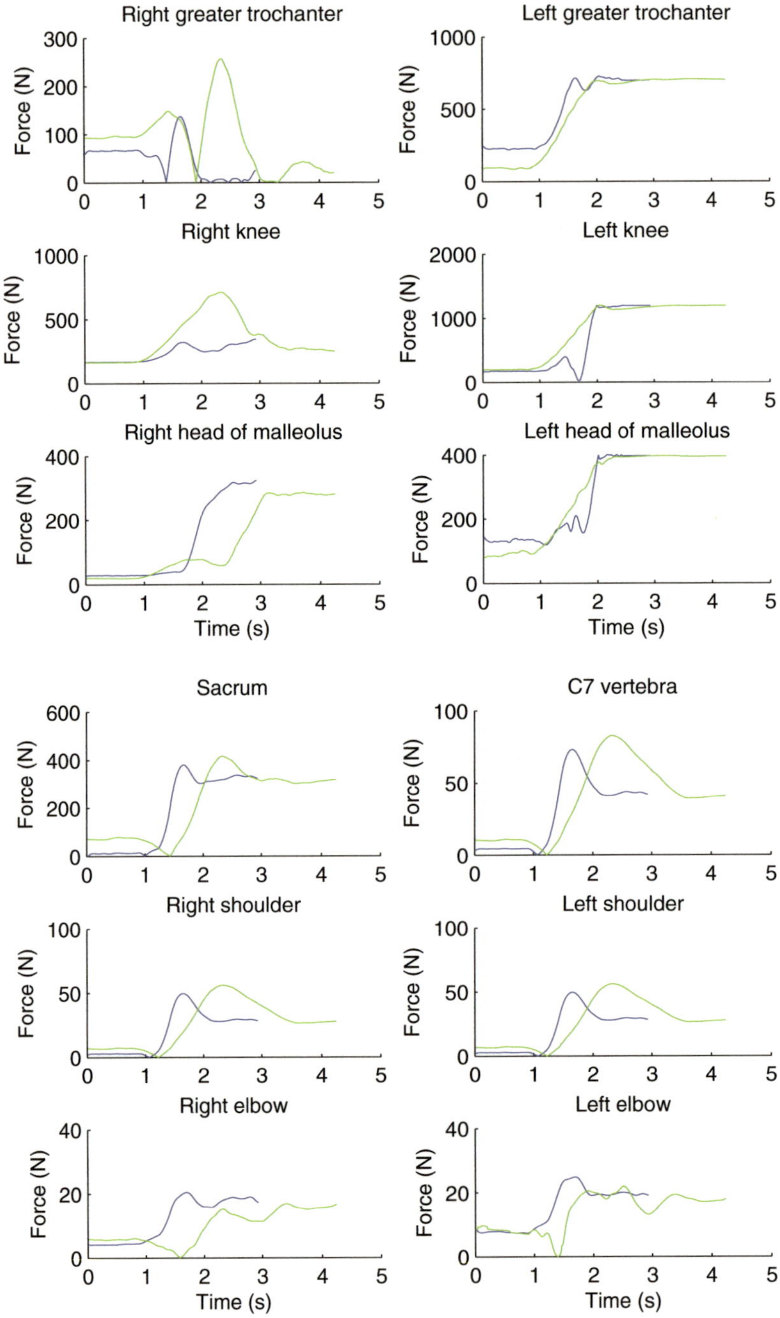

Figure 11. Time evolution of the joint forces: superposition of the three components of joint forces at each joint, during the slow (green) and fast (blue) getting-up motions.

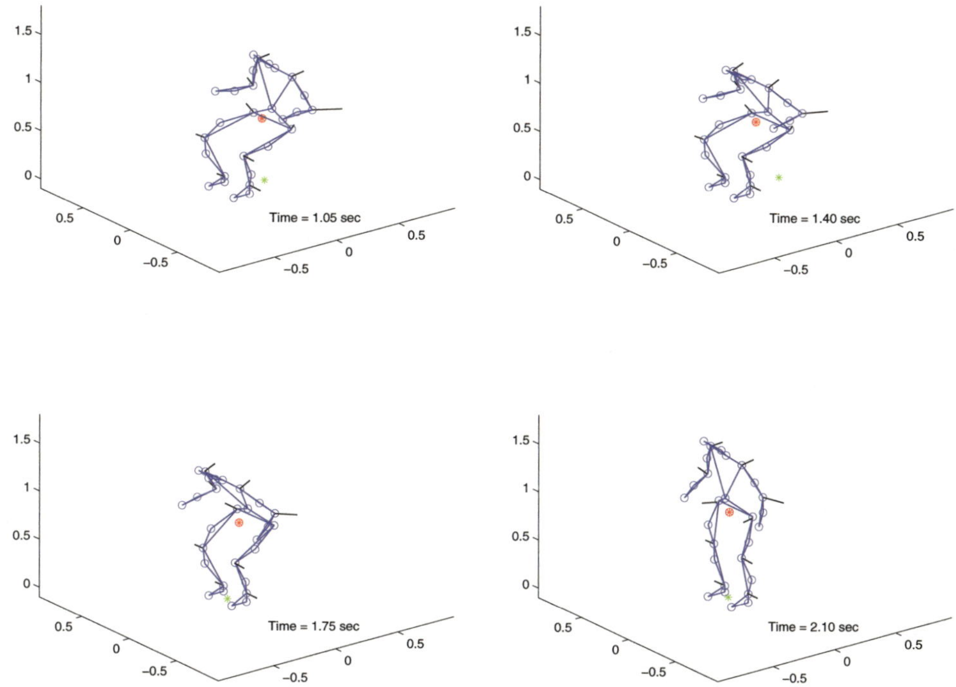

Figure 12. Samples of the fast STS, at four time instants t_k : the fitted model (in blue), featuring the global torques (in black) at each model joint, and also the GCOM (red point) and the GCOP (green star).

2. *The kinematic limitation, due to the rigid multibody system assumption :* Like other classical *dynamical inverse* analyzes [21–25] in biomechanics of motion, the proposed model is composed of linked rigid bodies. However, in reality, the body is not composed of a set of rigid bodies. Rather, each body member consists of a rigid part (bone), and a non-rigid part (skin, muscle, ligament, tendon, connective tissue, and other soft tissue structures) [38]: during any motion, the skeletal structures of the body experience accelerations, whereas the soft tissue motion is delayed, due to damped vibrations of the member. Consequently, the errors in the optimized joint coordinates q may introduce errors in the velocities $\dot{q}$ and accelerations $\ddot{q}$, and thus introduce errors in the estimation of the internal efforts.

3. *The dynamical limitation, due to the approximation of the body inertia parameters :* The body inertia parameters, i.e. the masses m_i, moments of inertia I_i and center of mass positions $\overrightarrow{OM_i}$ of the ith body member (i = 1,. . . ,13) are approximated, using inertia tables [31]. Consequently, the errors in the estimated internal efforts Q increase if the corresponding body member accelerations increase. This is the reason why the present model is only proposed for rather small dynamics, such as the STS experiment, walking experiments or other motions without significant impact. Further, let us remember that previous investigations [33] showed that the non-invasive body parameter identifications during the

motions are presently inappropriate to the human body dynamics, because the resulting body parameters present large errors due to experimental errors in the input data, such as the body configuration, external force and torque measurements.

4.2. Perspectives

Finally, in the context of the hardness to perform efforts, the perspectives of this research is to quantify with a satisfying accuracy the main joint and muscle efforts of subjects in different dynamical contexts, and to apply the model to:

- physical therapy, in order to analyze joint efforts of subjects in different motion contexts, particularly for the evaluation, the follow-up and the treatment of patients in rehabilitation and orthopedics;
- comfort analysis is vehicle and car occupant dynamics, in order to analyze the hardness of going into and out of vehicles, and simulate the car occupant dynamics before crash.

Acknowledgements

The research is supported by MÉDITIS training program, supported by NSERC/FONCER, Canada.

Author details

Raison Maxime, Laitenberger Maria and Sarcher Aurélie
Research Chair in Pediatric Rehabilitation Engineering (CPRE), École Polytechnique de Montréal and CRME - Sainte-Justine UHC, Montreal, Canada

Detrembleur Christine
Institute of NeuroScience (IoNS), Université catholique de Louvain (UCL), Brussels, Belgium

Samin Jean-Claude and Fisette Paul
Centre for Research in Mechatronics (CEREM), Institute of Mechanics, Materials, and Civil Engineering (iMMC), Université catholique de Louvain (UCL), Louvain-la-Neuve, Belgium

5. References

[1] Bohannon RW (2011) Test-retest reliability of the five-repetition sit-to-stand test: a systematic review of the literature involving adults. J Strength Cond Res 25(11): 3205-7.
[2] Dos Santos AN, Pavão SL, Rocha NA (2011) Sit-to-stand movement in children with cerebral palsy: a critical review. Res Dev Disabil 32(6): 2243-52.
[3] Wang TH, Liao HF, Peng YC (2011) Reliability and validity of the five-repetition sit-to-stand test for children with cerebral palsy. Clin Rehabil. Epub ahead of print.
[4] Duncan RP, Leddy AL, Earhart GM (2011) Five times sit-to-stand test performance in Parkinson's disease. Arch Phys Med Rehabil 92(9): 1431-6.
[5] Jovic J, Fraisse P, Coste CA, Bonnet V, Fattal C (2011) Improving valid and deficient body segment coordination to improve FES-assisted sit-to-stand in paraplegic subjects. IEEE Int Conf Rehabil Robot. Jun 29 - Jul 1.

[6] Wetzel JL, Fry DK, Pfalzer LA (2011) Six-minute walk test for persons with mild or moderate disability from multiple sclerosis: performance and explanatory factors. Physiother Can 63(2): 166-80.

[7] Gao F, Zhang F, Huang H (2011) Investigation of sit-to-stand and stand-to-sit in an above knee amputee. Conf Proc IEEE Eng Med Biol Soc: 7340-3.

[8] Agrawal V, Gailey R, Gaunaurd I, Gailey R 3rd, O'Toole C (2011) Weight distribution symmetry during the sit-to-stand movement of unilateral transtibial amputees. Ergonomics 54(7): 656-64.

[9] Noguchi H, Hoshiyama M, Tagawa Y (2012) Kinematic analysis of sit to stand by persons with rheumatoid arthritis supported by a service dog. Disabil Rehabil Assist Technol 7(1): 45-54.

[10] Turcot K, Armand S, Fritschy D, Hoffmeyer P, Suvà D (2012) Sit-to-stand alterations in advanced knee osteoarthritis. Gait Posture. Epub ahead of print.

[11] Boyne P, Israel S, Dunning K (2011) Speed-dependent body weight supported sit-to-stand training in chronic stroke: a case series. J Neurol Phys Ther 35(4): 178-84.

[12] Kim I, Cho W, Yuk G, Yang H, Jo BR, Min BH (2011) Kinematic analysis of sit-to-stand assistive device for the elderly and disabled. IEEE Int Conf Rehabil Robot. Jun 29 - Jul 1.

[13] Jun HG, Chang YY, Dan BJ, Jo BR, Min BH, Yang H, Song WK, Kim J (2011) Walking and sit-to-stand support system for elderly and disabled. IEEE Int Conf Rehabil Robot. Jun 29 - Jul 1.

[14] Kinoshita S (2012) Handrail position and shape that best facilitate sit-to-stand movement. J Back Musculoskelet Rehabil 25(1): 33-45.

[15] Silva M, J Ambrósio, Pereira M (1997) A multibody approach to the vehicle and occupant integrated simulation. International Journal of Crashworthiness 2(1): 73-90.

[16] Pérez M, Ausejo S, Pargada J, Suescun A, Celigüeta JT (2003) Application of multibody system analysis for the evaluation of the driver's discomfort. Proceedings CD-rom of the multibody dynamics, July 1-4, Lisbon, Portugal.

[17] Bouisset S, Le Bozec S, Ribreau C (2002) Postural dynamics in maximal isometric ramp efforts. Biological Cybernetics 87(3): 211-19.

[18] Raison M, Detrembleur C, Fisette P, Willems PY (2004) Determination of joint efforts of a moving human body by inverse dynamics. Archives of Physiology and Biochemistry 112, suppl. September, 1-179: 90.

[19] Fujimoto M, Chou LS (2012) Dynamic balance control during sit-to-stand movement: an examination with the center of mass acceleration. J Biomech 45(3): 543-8.

[20] Raison M, Detrembleur C, Fisette P, Samin JC (2011) Assessment of Antagonistic Muscle Forces During Forearm Flexion/Extension. Multibody Dynamics: Computational Methods and Applications 23: 215-38.

[21] Denoth J, Gruber K, Ruder H, Keppler M (1984) Forces and torques during sport activities with high accelerations. Biomechanics current interdisciplinary research. Eds. Perren SM and Schneider E. Martinuis Nijhoff Pub. Dodrecht, Netherlands: 663-668.

[22] De Jalón G, Bayo E (1993) Kinematic and dynamic simulation of multibody systems: the real-time challenge. Springer, New-York: 440 p.

[23] Silva M, Ambrósio J, Pereira M (1997) Biomechanical Model with Joint Resistance for Impact Simulation. Multibody System Dynamics 1(1): 65-84.

[24] Silva M, Ambrósio J (2004) Sensitivity of the Results Produced by the Inverse Dynamic Analysis of a Human Stride to Perturbed Input Data. Gait and Posture 19(1): 35-49.

[25] Zacher I (2004) Strength Based Discomfort Model of Posture and Movement. SAE International Digital Human Modelling Conference, June 15-17, Rochester, Michigan.
[26] Fisette P, Postiau T, Sass L, Samin JC (2002) Fully symbolic generation of complex multibody models. Mechanics of Structures and Machines 30(1): 31-82.
[27] Samin JC, Fisette P (2003) Symbolic Modeling of Multibody Systems, Kluwer Academic Publisher: 484 p.
[28] Detrembleur C, Van Den Hecke A, Dierick F (2000) Motion of the body centre of gravity as a summary indicator of the mechanics of human pathological gait. Gait and Posture 12: 243-50.
[29] Bouisset S, Maton B (1996) Muscles, Posture et Mouvement. Herman Edition, Paris, 735 p.
[30] Bouisset S (2002) Biomécanique et physiologie du mouvement. Abrégés, Editions Masson: 304 p.
[31] De Leva P (1996) Adjustments to zatsiorsky-seluyanov's segment inertia parameters. Journal of Biomechanics 29(9): 1223-30.
[32] Zatsiorsky VM, Seluyanov VN, Chugunova L (1990) In vivo body segment inertial parameters determination using a gamma-scanner method. Biomechanics of human movement: Applications in rehabilitation, sports and ergonomics; edited by Berme N and A Cappozzo: 187-202.
[33] Chenut X, Fisette P, Samin JC (2002) Recursive Formalism with a Minimal Dynamic Parametrization for the Identification and Simulation of Multibody Systems. Application to the Human Body. Multibody System Dynamics 8: 117-40.
[34] Bao H, Willems PY (1999) On the kinematic modelling and the parameter estimation of the human knee joint. Journal of Biomechanical Engineering 121: 406-13.
[35] Bao H, Willems PY (1999) On the kinematic modelling and the parameter estimation of the human shoulder. Journal of Biomechanics 32: 943-50.
[36] Heglund N (1981) A simple design for a force-plate to measure ground reaction forces. Journal of Experimental Biology 93: 333-38.
[37] Eames MHA, Cosgrove A, Baker R (1998) A Full body model to determine the total body centre of mass during the gait cycle in adults and children. Fifth International Symposium on the 3-D Analysis of Human Movement, Chattanooga, Tennessee, USA, July 2-5.
[38] Nigg BM, Herzog W (1999) Biomechanics of the musculo-skeletal system. Eds. Nigg BM and W Herzog. 2d edition. Chichester and New York. 644 p.
[39] Raison M, Aubin CE, Detrembleur C, Fisette P, Mahaudens P, Samin JC (2010) Quantification of intervertebral efforts during gait: comparison between subjects with different scoliosis severities. Studies in Health Technology and Informatics 158: 107-11.
[40] Raison M et al. Website of the Research Chair in Pediatric Rehabilitation Engineering, École Polytechnique de Montréal and CRME - Sainte-Justine UHC, Montreal, Canada. Available: `www.groupes.polymtl.ca/cgrp`. Accessed 2015 Apr 15.

Chapter 9

Modeling the Foot-Strike Event in Running Fatigue via Mechanical Impedances

J. Mizrahi and D. Daily

Additional information is available at the end of the chapter

http://dx.doi.org/10.5772/48784

1. Introduction

The human motor system benefits from remarkable muscular redundancies: A motor task is normally performed with the simultaneous involvement of more muscles than strictly necessary. Furthermore, this same task may be performed in multiple ways, with different muscle combinations. From the mechanical viewpoint the musculoskeletal system is indeterminate, whereby the number of unknown muscle forces exceeds the number of available equations. We address in this Chapter the biomechanics of the lower limbs in long-distance running under conditions of developing fatigue. In long-distance running the running speed may result in more than 300 foot-strikes per leg per kilometer. Each such foot-strike evokes an impact loading that results in a vertical shock impulse transmitted upwards through the body and carries with it the potential for injuries in the bone and joint tissues.

Fatigue, or stress, fractures occur in bones in response to repetitive stresses over multiple cycles, when the body's ability to adapt is exceeded [1,2]. An important factor which affects the incidence of bone stress injury, is exposure to abrupt changes in the bone loading [1], and consequent alteration in the strain distribution [3] with insufficient recovery periods [4]. Other factors include footwear, terrain and intensity of activity or training [1].

Two of the major factors responsible for impulse attenuation at foot- or heel-strike are the shock absorption capacity of the active muscle in the lower limbs, and the cushioning effect of the foot heel-pad tissue. In previous reports we have shown that in long distance running the impact shock load on the lower limbs increases with progressing fatigue [5-8]. One additional question is whether, as a result of fatigue, an imbalance between the activities of the plantar and dorsi flexor muscles of the ankle develops. Such an imbalance would compromise the protective action provided by the muscles to the shank [9-11].

The goal of this research was to characterize the heel-strike shock propagation and attenuation in running by means of a biomechanical model, and to examine changes taking place as a result of running fatigue.

2. Biomechanical modeling of the lower limb

This section deals with the modeling of the heel-strike event. With the development of biomechanical models of human body motion, it has become possible to simulate vertical landing, such as occurring during running, in order to gain insight into intermuscular coordination and to elucidate control strategies of the musculoskeletal system. A common method to deal with this type of problems is to lump together elements of the human body e.g., muscles, tendons, ligaments, bones and joints so that the overall musculoskeletal system is represented as a damped elastic mechanism. Several models describing vertical landing can be found in the literature [12-18]. These models are usually characterized by the presence of elastic springs and viscous dampers, with constant properties and provide a reasonable prediction of the maximal vertical foot/ground reaction force.

Indeterminacy of the locomotor system can be addressed by adopting the lumping method, whereby the material elements of the human body e.g., muscles, tendons, ligaments, bones and joints are lumped together so that the overall musculoskeletal system is represented as a multi-degree-of freedom damped elastic mechanism, interconnecting the masses of the body segments.

The foot- or heel-strike period during landing from fall, during hopping or during the stance phase of running has been generally modeled using one-dimensional models along the vertical direction. [13-15,19-20].

In this study we represent the body segments during heel-strike by a four degree-of-freedom elastically-damped uni-axial biomechanical model. The model thus includes 4 masses connected by elastic stiffnesses with parallel damping elements, as shown in Figure 1. In more details, the masses m_j (j = 1..4) represent, respectively, the foot + shoe, the shank, the thigh and the rest of the body (including the non-supporting leg). Each of the stiffness k_j and damping c_j (j = 1..4) represent, respectively, the lumped effects of the heel-pad + sole, the ankle joint, the knee joint and the hip joint.

3. Model equations

For the above model, the force diagram, as presented in Figure 2, yields the following model equations.

$$\begin{cases} m_1\ddot{x}_1 = -k_1x_1 - c_1\dot{x}_1 - k_2(x_1 - x_2) - c_2(\dot{x}_1 - \dot{x}_2) - m_1\mathbf{g} \\ m_2\ddot{x}_2 = k_2(x_1 - x_2) + c_2(\dot{x}_1 - \dot{x}_2) - k_3(x_2 - x_3) - c_3(\dot{x}_2 - \dot{x}_3) - m_2\mathbf{g} \\ m_3\ddot{x}_3 = k_3(x_2 - x_3) + c_3(\dot{x}_2 - \dot{x}_3) - k_4(x_3 - x_4) - c_4(\dot{x}_3 - \dot{x}_4) - m_3\mathbf{g} \\ m_4\ddot{x}_4 = k_4(x_3 - x_4) + c_4(\dot{x}_3 - \dot{x}_4) - m_4\mathbf{g} \end{cases} \tag{1}$$

with initial conditions:

$$x_i(0)=0 \;\; ; \;\; \dot{x}_i(0)=v_0=-1\ m/s \;\; , \;\; i=1,2,3,4 \tag{2}$$

and gravitational acceleration

$$\mathbf{g}=-9.81\ m/s^2$$

These values rely on reported landing velocities between -0.8 m/s to -1.2 m/s for running speeds of 3.5 m/s (comparable to the speeds of this study), while wearing various types of running shoes [21-23].

The above masses are expressible in terms of the total body mass from anthropometric data [24].

$$\begin{cases} m_1 = 0.0145 \cdot m \\ m_2 = 0.0465 \cdot m \\ m_3 = 0.100 \cdot m \\ m_4 = 1-(m_1+m_2+m_3) = 0.839 \cdot m \end{cases} \tag{3}$$

From the simultaneous recording of the foot ground reaction forces and accelerations on the masses m_j, information about the rise time of the peak acceleration can be obtained.

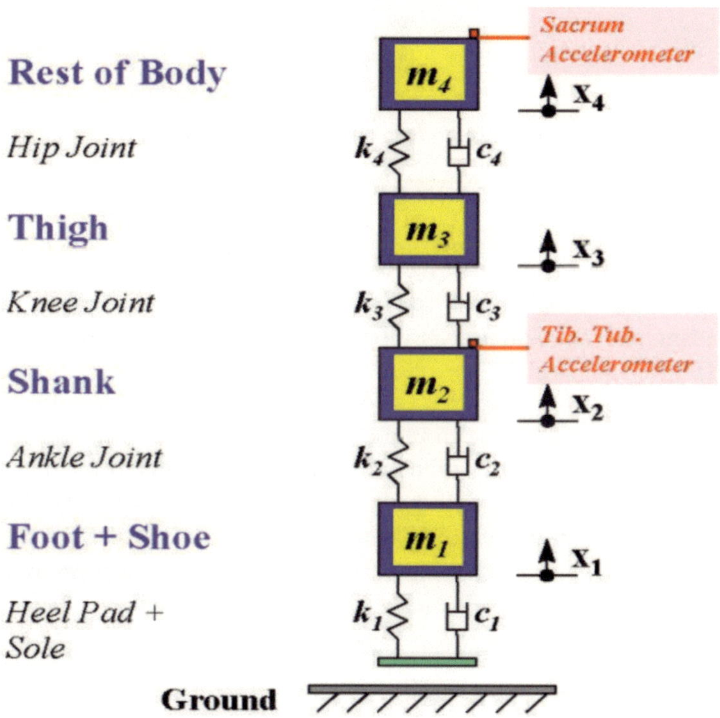

Figure 1. Lumped model including 4 masses connected by elastic stiffnesses with parallel dampings. The masses m_j (j = 1..4) represent, respectively, the foot + shoe, the shank, the thigh and the rest of the body (including the non-supporting leg). Each of the stiffness k_j and damping c_j (j = 1..4) represent, respectively, the lumped effects of the heel-pad + sole, the ankle joint, the knee joint and the hip joint.

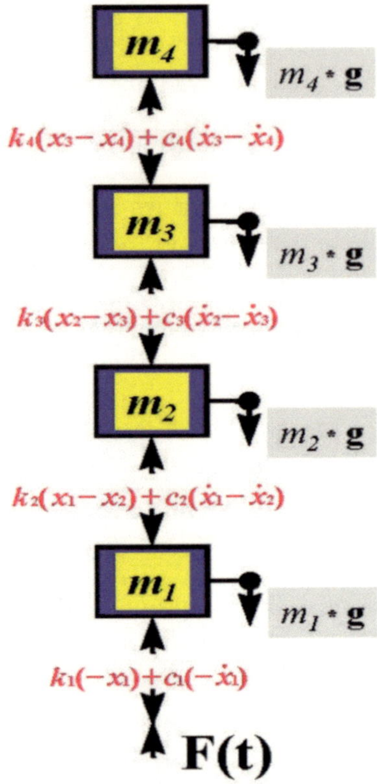

Figure 2. The force diagram for the model elements presented in Figure 1.

4. Experimental setup

Information about the impulsive loading along the skeletal elements in long-distance running can be non-invasively obtained from the foot-ground reactive forces [25] and, more directly, by measuring the transient accelerations on the shank caused by impact.

4.1. Impact accelerations

Non-invasive in vivo measurements of acceleration and impact transmission along the human body were made by externally attaching light-weight, high-sensitivity accelerometers at strategic points including bony prominences, such as the tibial tuberosity below the knee area, the greater trochanter near hip level and the sacrum area at the lower back [13,15,22,26-29].

In this study, each subject was instrumented with two light-weight (4.2 grams) uniaxial (Kistler PiezoBeam, type 8634B50, Kistler, Winterthur, Switzerland), skin-mounted accelerometers connected to a coupler (Kistler Piezotron, type 5122). One was attached on the tibial tuberosity, and the second - on the sacrum. To achieve good reliability of the measurements by means of bone-mounted accelerometers, the accelerometers were pressed onto the skin in closest position to the bony prominences of the tibial tuberosity and the sacrum, by means of two elastic belts passed in a horizontal plane around the shank and the waist, respectively. The tensions of the belts were well above the level in which the acceleration trace for a given impact force became insensitive to the accelerometer attachment force, thus ensuring stability of the accelerometer as well as consistency of the readings and reproducibility of the data [13,30].

The shank accelerometer was aligned with the axis of the tibia to provide the axial component of the tibial acceleration and the accelerometer on the sacrum was oriented along the spine. These accelerometers allowed us to acquire the shock accelerations propagated in the longitudinal directions of the tibia and the spine. As earlier reported, such attachment is suitable for faithfully measuring the amplitude of shock acceleration [5-8].

Force platforms, type Kistler Z-4035, were used for the simultaneous recordings of the foot-ground reaction forces and acceleration.

4.2. Running fatigue tests

An overview of the experimental setup is shown in Figure 3. For examining the effect of global fatigue due to running, the subjects were asked to run on a Quinton Q55 treadmill.

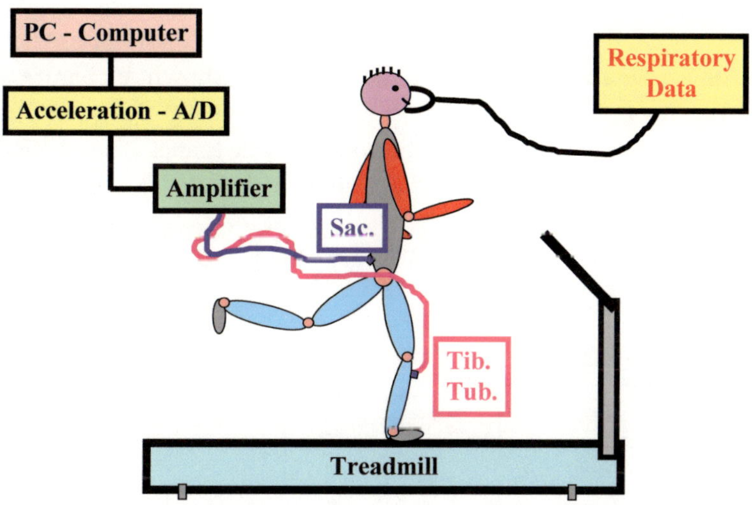

Figure 3. Description of the experimental apparatus with a subject running on a treadmill. Two accelerometers are attached , one on the tibial tuberosity (Tib. Tub.) and the other on the sacrum (Sac.) ; the accelerometers' data are sampled through an amplifier and A/D converter to a PC. Likewise, the respiratory data are sampled and collected on-line.

Global, or metabolic fatigue is associated with the development of metabolic acidosis following an endurance exercise and is accompanied by a decrease in the end tidal carbon dioxide pressure (PETCO$_2$) [31]. In long distance running metabolic fatigue is reached when the running speed exceeds the anaerobic threshold [31].

Running was thus for a duration of 30 min and at a speed exceeding the anaerobic threshold speed of each subject by 5%. Before the test a 15 min warming up running on the treadmill was performed. In this study, the average running speed for all 14 subjects was 3.53 m/s (SD, 0.19). It should be noted, however, that in addition to global fatigue, local fatigue in a muscle may also take place as a result of an intensive activity of this muscle. This type of fatigue is reflected by certain changes in its electromyogram (EMG) signal in the time and/or frequency domains [32]. Local fatigue was not considered in this study.

Respiratory data were collected from a Sensor-Medics 4400 device and included $\dot{V}_{O_2}$ - Oxygen consumption, $\dot{V}_{CO_2}$ - CO_2 production, $\dot{V}_E$ - minute ventilation, $PETCO_2$ - end-tidal CO_2 pressure, $\dot{V}E / \dot{V}O_2$ - ventilatory equivalent for oxygen, $\dot{V}E / \dot{V}CO_2$ - ventilatory equivalent for CO_2. The anaerobic threshold was determined as the point of initial increase of $\dot{V}E / \dot{V}O_2$ and $\dot{V}E / \dot{V}CO_2$, which just precedes the initial decline of $PETCO_2$ [31]. In a previous study, we have shown that 30 min running at a speed exceeding the anaerobic threshold is a sufficient time to induce general fatigue [8].

The respiratory data were evaluated at each of the 1st, 5th, 10th, 15th, 20th, 25th and 30th min of running and the accelerometer and force platform data were online sampled at 1667 Hz sampling rate. The model parameters were estimated, however, at the 1st, 15th and 30th min of running.

The dynamics of acceleration build-up at heel-strike is shown in Figure 4 where the simultaneous recordings of the tibial tuberosity acceleration and the ground reaction force (GRF) are shown in two time scales: complete running cycle (panel a) and zooming-in on the heel-strike event (panel b). In this case the build-up time to the tibial tuberosity peak acceleration was ~ 30 ms. It is also noted that the ground reaction force exhibits two peaks: a smaller one shortly after heel-strike and a larger one (~ 2.5 body weights), towards the middle of the running cycle.

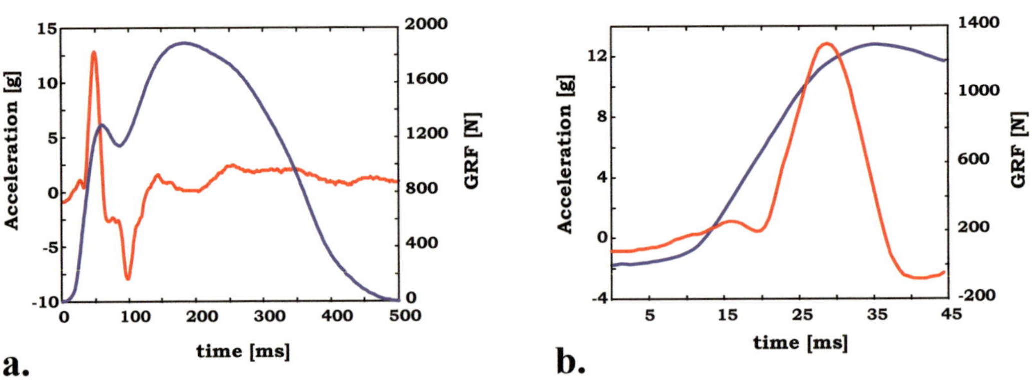

Figure 4. Typical foot-ground reaction force (GRF) (transient curve) measured simultaneously with tibial tuberosity acceleration (spiky curve). **a.** complete stance phase; **b.** heel-strike period only.

Typical accelerometer traces for the tibial tuberosity (right leg) and sacrum are shown in Figure 5, for a complete running cycle, i.e., from heel-strike of the right foot till the next heel-strike of the same foot. Two major differences should be noted between the traces: (a) intensity of ~ 8 g in the tibial tuberosity versus less than 3 g in the sacrum; (b) while the tibial tuberosity exhibits one major peak within the first 50 ms of the running cycle from heel-strike and originating from the heel-strike of the right foot, the sacrum acceleration, due to its central location, exhibits two comparable positive peaks within the running cycle, reflecting each of the right and left heel-strikes.

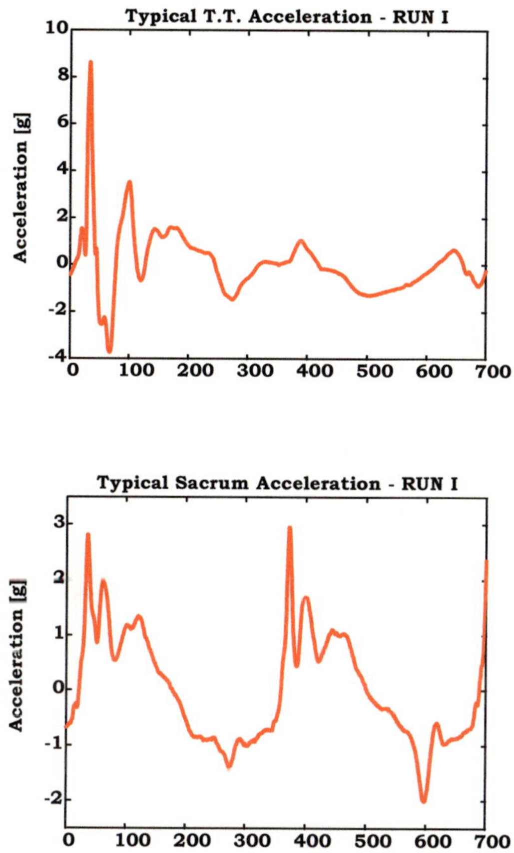

Figure 5. Typical accelerometer traces for the tibial tuberosity (T.T.) (right leg) and sacrum, for a complete running cycle, i.e., from heel-strike of the right foot till the next heel-strike of the same foot.

5. Parameter estimation of the model

The mechanical properties of biological material are, in general, multiple variable-dependent. Specifically stiffness, in addition to its being non-linear e.g. strain dependent, often depends on the deformation rate. This is also the case with bones [33], tendons and ligaments [34], cartilage [35] and muscle [36]. Damping too may be position-dependent.

Due to nonlinearity of the stiffness/damping properties of the joints of the leg [e.g. 20,37], we were not generally able to estimate the model parameters while assuming that they remain constant over the heel-strike period. Thus, the heel-strike period was divided into two equal periods (22 ms each) and the parameters were estimated separately for each of these periods, using the Gauss-Marquardt [38-39] method of non-linear estimation. For the first period the initial conditions were as prescribed in equation (2), and for the second period the initial conditions used were the values reached at the end of the first period.

Figure 6 shows the model prediction of the shank mass (m_2 in Figure 1) acceleration (continuous line) versus the experimentally measured tibial tuberosity acceleration (denoted as +). The results of two different subjects are demonstrated (vertical partition of the Figure) for two time points of running (horizontal partition): 1st min, fatigue-free, and 15th min of the running test. The subtle, brief, discontinuity in the traces at mid-cycle represents the transition between the two parameter estimation periods.

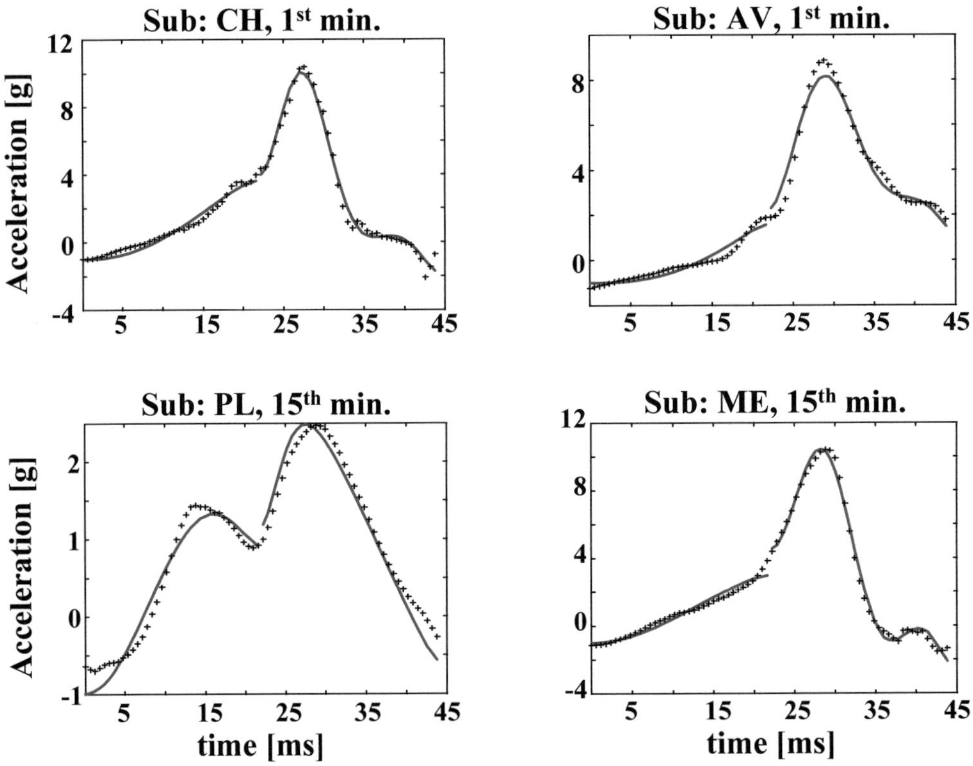

Figure 6. The model prediction of the m_2 acceleration (continuous line) versus the experimentally measured tibial tuberosity acceleration (denoted as +). The upper panels represent the solution for the 1st min of running and the lower panels represent the solution for the 15th min of running. The subtle, short, discontinuity in the traces at mid-cycle represents the transition between the two parameter estimation periods, with piece-wise constant stiffness each. Each of the left and right panels represents a different subject.

6. Model results and model sensitivity analysis

Table 1 shows the sample results of the stiffness and damping parameters for the two time zones. A sensitivity analysis can provide an indication to the quality of the estimation of parameters. Thus, sensitivity of the m_2 and m_4 model results to each of the estimated parameters was performed by two-fold multiplying and/or dividing each of the estimated parameters separately and for each time zone. The two-fold variation demonstrated a strong sensitivity of m_2 to k_1 and k_2 in the first time zone and to parameters k_1, k_2, k_3 and c_3 in the second time zone. Small to medium sensitivity was obtained in the second time zone to parameters c_1 and k_4, repectively. The m_4 acceleration was not sensitive in the first time zone to none of the parameters, while in the second time zone it demonstrated strong sensitivity to parameters k_2 and k_3 and small sensitivity to parameters k_4 and c_3.

Tested Parameter k_i [N/m] c_i [NS/m]	Initial parameter value		sensitivity m_4 acceleration		sensitivity m_2 acceleration	
	ZONE 1	ZONE 2	ZONE 1	ZONE 2	ZONE 1	ZONE 2
k_1	16720	265305	***	***	* (**)	* (***)
k_2	14276	31563	***	***	* (*)	***
k_3	3946	16427	* (**)	***	* (*)	***
k_4	150000	150170	* (*)	** (**)	* (*)	** (***)
c_1	34.4	250.6	* (***)	** (***)	* (**)	* (*)
c_2	1.1	1.1	* (*)	* (*)	* (*)	* (*)
c_3	1.2	488	* (*)	***	* (*)	** (**)
c_4	998	1000	* (*)	* (*)	* (*)	* (*)

Table 1. Summary of the sensitivity of the m_2 and m_4 accelerations to the model parameters. A twofold change up and down was checked in the two time solution-zones, separately for each parameter. Parameters for which the accelerations were not sensitive were checked also for a change in one order of magnitude (up and down). The results of this latter test are shown in parentheses (* = negligibly low effect; ** = low to medium effect; *** = high effect, i.e. more than 10% change in peak acceleration of m_2 or m_4).

In cases of low- or no-sensitivity the parameters were varied, up and down, by one order of magnitude with the results shown in parentheses. The one order of magnitude variation in the parameter values did not evoke sensitivity beyond that of the twofold variation, except for c_1 on m_2 and k_1 on m_4.

Due to the fact that the parameter estimation of the model coefficients was performed in short time intervals, fractions of the heel strike event, the damping coefficients disclosed a high variability. Better estimations would have probably resulted if the period of estimation was higher. Accordingly, the stiffness parameters k_1 and k_2, in the first time zone, and k_1, k_2 and k_3 in the second time zone, considered more reliably estimated, were reported in what follows.

Difficulties in estimating the damping coefficients are not unusual due to their expected low values. It has been reported that in repetitive physical activity, such as in running, the subject bounces on the ground in a spring-like manner [17, 40-45]. Depending on the range of joint flexion and on the frequency of motion, a considerable amount of elastic energy can be stored and re-used. It has been shown that the dissipated energy in muscles increase when the amplitudes of joint movement are increased [46]. It has also been also commented that the utilization of stored elastic energy depended on the shortness in latency between the stretch and shortening phases of the muscles [47]. Accordingly, during the ground-contact period of running, the leg was modeled as a one-dimensional four-degree-of-freedom piece-wise linear spring, with no damping. During heel-strike, the joints did not have a damping effect, to contribute to energy dissipation.

Summary of the values of these parameters for the two solution time zones is given in the rightmost column of Table 2. The values, averages for 10 subjects (SD), are calculated at each of the 1st, 15th and 30th min of running to evaluate the effect of fatigue. The asterisks indicate a significant change $p < 0.05$ between the values of zone 1 and zone 2. The differences between k_1 in zone 1 and k_1 in zone 2 were significant with $p < 0.0007$, for each of the 1st, 15th and 30th min of running. For the k_2 parameter a statistically significant difference was obtained for the 1st min of running only, with $p < 0.009$. The differences in parameter values due to fatigue were not statistically different, despite the notable differences in the averages. The reason was the big variability among the tested subjects. On the individual level, though, the differences due to fatigue were statistically significant in most of the cases.

It should be noted that the stiffness k_1, relating to the heel-pad may be alternatively obtained from the ratio between the foot-ground reaction force and the heel-pad deformation. Past reports using this method have reported an increased stiffness with the heel-pad deformation, as occurs during heel-strike [27,48-52]. These studies have reported an approximately tenfold increase in heel-pad stiffness, i.e., from 6.6 – 135 kN/m to 77 – 1430 kN/m, using different methodologies including actual running in conjunction with measurements of the heel pad by means of X-rays; heel-strike simulation by in vivo pendulum impact, or Instron measurements made on cadavers. Despite the differences with the method applied in this study, the stiffness k_1 indicates similar values with those obtained by other methods. Our model provides further the fatigue effect on k_1, which is found opposite in the first time zone, where the stiffness decreases with fatigue, to the second time zone where the stiffness is found to increase.

	Stiffness [kN/m]	Time in running [min]		
		1	15	30
ZONE 1	k_1	19.3 (28.2)	16.4 (16.1)	13.6 (4.7)
	k_2	11.4 (4.3)	14.7 (6.0)	19.8 (18.0)
ZONE 2	k_1	140.2* (90.1)	190.3* (53.8)	171.0* (73.2)
	k_2	17.1* (4.5)	17.8 (4.4)	23.1 (15.6)
	k_3	14.0 (9.3)	17.8 (7.3)	15.8 (9.4)

Table 2. Parameter means (S.D) of 10 subjects: k_1, k_2 in ZONE 1 (upper part of Table), and k_1, k_2, k_3 in ZONE 2 (lower part of Table). Parameter-values are from 3 time points, i.e. at the 1st, 15th and 30th min of running. Parameter-means change with fatigue was significant with $p>0.148$. The asterisks in the lower part of Table signify significant changes ($p<0.05$) from ZONE 1 to ZONE 2.

The m_1 displacement obtained from our results revealed that the foot sinks about one cm in the first time zone of heel-strike, followed by a bouncing back phase. Previous reports also indicated a 1 cm deformation of the heel [e.g. 48]. The elastic properties of the heel-pad of various mammals were studied and it has been found that full up and down oscillations might result before actual settling down of the foot [53]. At cases, the foot may even temporarily lose contact with the ground during these oscillations. In our case, the oscillation phase would succeed the first time zone.

The obtained average knee stiffness k_3 of 15.8 kN/m was within the range of 6.89 – 112.0 kN/m, previously reported [12-14,16,19,22,29,45,54-55].

Further exploration of the effect of fatigue in the course of running was performed by correlating, using linear regression, the tibial tuberosity peak acceleration to the parameter values obtained from the model results. Figure 7 shows the correlation for which the Pearson's coefficient was statistically different from zero (with 95% significance). The parameters are k_1 in the first and second zones, upper and middle rows, respectively, and k_3 in the second time zone of stance. The 1st, 15th and 30th min of running are shown in the left, middle and right columns, respectively. Table 3 summarizes the regression results for the cases displayed in Figure 7. The results show the following: (a) For k_1 in time zone 1, the Pearson correlation r_p starts off by a low value of 0.26 in the 1st min of running but increases in the 15th (0.49) and the 30th (0.44) min of running, indicating that a higher peak in the tibial tuberosity acceleration is associated with a lower k_2 in time zone 1; (b) there is a high correlation (0.86) between the peak tibial tuberosity acceleration and the k_1 parameter in time zone 2 at the initial stage of running (1st min).

This correlation, however, was lower in the 15th min (0.27) and in the 30th min (0.49) of running, thus suggesting that high peak acceleration at the tibial tuberosity is associated with a higher k_1 value in time zone 2; (c) a medium correlation (0.73) was found between k_3 and peak acceleration at the tibial tuberosity in the 1st mi of running. But here too this correlation was decreased with the development of fatigue, suggesting that higher peak acceleration at the tibial tuberosity is associated with a high k_3 parameter value in time zone 2.

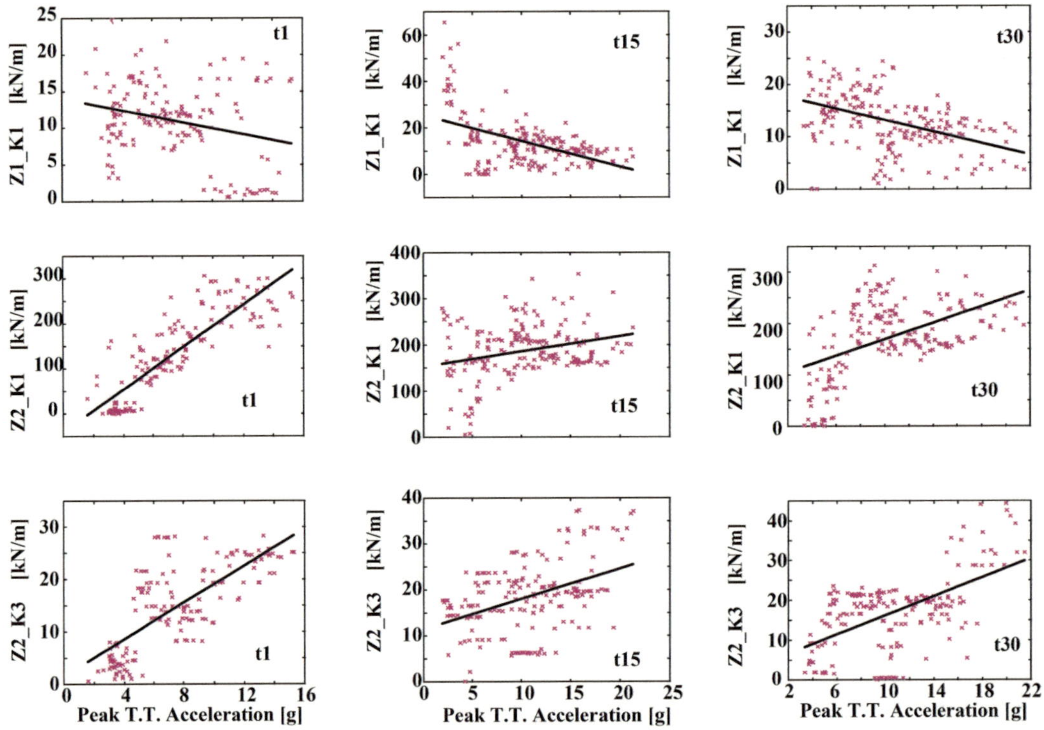

Figure 7. Simple linear-regression made to express the relationship between the tibial tuberosity peak acceleration and parameter values (results of 10 subjects). The regression was performed for the 1st min (leftmost panels), the 15th min (middle panels) and for the 30th min (rightmost panels). The correlation coefficient of the presented regression lines was significantly different from zero, at 95% significance level. The 3 upper panels - for k_1 in ZONE 1 (Z1); 3 middle panels for k_1 in ZONE 2 (Z2); and 3 lower panels - for k_3 in ZONE 2.

It has been shown that in running with shoes, the foot is restricted from bulging sideways, thus limiting the vertical deformation to an average of 5.5 mm, as opposed to 9 mm when running barefoot [48,56]. This explains the higher stiffness during the first time zone compared to bare foot running. It also explains the lower stiffness during the second time zone compared to bare foot running. It has also been shown by that better energy absorption and impact shock attenuation is associated with lower stiffness [51].

Tested parameter	Regression coefficients	t1	t15	t30
Z1_k_1	a	-0.41±0.23	-1.10±0.29	-0.56±0.16
	b	14.03±1.86	25.23±3.27	18.77±1.85
	r_p	0.26	0.49	0.44
	r^2	0.07	0.24	0.19
Z2_ k_1	a	23.56±2.20	3.26±1.70	7.94±2.02
	b	40.36±17.92-	153.10±19.48	89.95±23.05
	r_p	0.86	0.27	0.49
	r^2	0.74	0.07	0.24
Z2_ k_3	a	1.76±0.26	0.66±0.20	1.19±0.24
	b	1.52±2.08	11.45±2.34	4.26±2.74
	r_p	0.73	0.42	0.57
	r^2	0.54	0.18	0.33

Table 3. Linear regression results (Y= aX+b) of Figure 7, at 95% confidence intervals for coefficients a and b ; r_p = Pearson correlation coefficient ; and r^2 = coefficient of determination.

The correlation found between low stiffness in the first time zone to the high stiffness in the second time zone is obvious from the anatomy of the heel pad, which consists of nearly closed collagen cells filled with fatty cells [27,48]. The vertical orientations of these cells, together with the high viscosity of the fat tissue are the major factors responsible for the absorption of the impact energy at heel strike. Initially, the fat flows sideways and small loads result in high deformation (low stiffness). In the second time zone, after the heel pad has already considerably deformed, further increase in deformation provokes a high load, thus high stiffness. The effect of fatigue could be explained by means of the heating effect during the course of running. With nearly 80 heel-strikes per min, the whole running duration of 30 min results in some 2500 heel-strikes, each of which causing a rapid deformation of the heel pad and during which the fatty tissue frictions while squeezed out of the collagen cells. The accumulated heat due to friction reduces the viscosity and the vertical displacement is accelerated, causing a reduction in stiffness during the first zone of heel strike. In the second zone, however, the thinner remaining tissue together with the underlying bone evokes an increase in stiffness.

7. Conclusion

Stress fractures in long bones of the lower limbs are believed to originate from repetitive and/or excessive loading, such as may take place in long-distance running at a speed exceeding the anaerobic threshold. In the present study the average running distance per test was 6.30 km (30 min of running at the average speed of 12.6 km/h) in agreement with the definition of 'long distance' [57]. We have measured and analyzed the following: respiratory data to monitor global fatigue; and accelerometry, to provide quantitative information on loading of the major segments of the lower limb. While providing accelerometer boundary values for the model system, accelerometry is an advantageous method due to its being non-invasive.

We have addressed a major fatigue-related factor taking part in exposing the shank to stress fractures risk: the decline in end tidal carbon dioxide pressure, the latter expressing metabolic fatigue [31,58]. The mechanical consequence of fatigue in long-distance running is two-fold: enhanced impact acceleration due to global fatigue and muscle activity imbalance due to local fatigue before and during foot contact, resulting in the development of excessive shank-bone bending stresses and higher risk of stress injury [11].

While departing from the stiffness constancy concept, the model revealed that a correct and sufficient variability of the joint stiffness is a two-region piece-wise constant stiffness indicating that a higher order of non-linearity is not necessary. This result should be considered meaningful in those problems where the constant stiffness representation is not sufficient and in cases where the system's representation has to be improved. Joint stiffness is dominated by muscular activation [59-60] and as the joints stiffen, they undergo smaller angular displacements during the ground-contact phase, also resulting in smaller excursion of the hip and higher leg stiffness. Thus, since stiffness is related to muscle activation, the piece-wise constant stiffness obtained solution also provides, through the obtained stiffness profiles, an insight into the patterns of the muscular activation in the legs' joints.

The fact that the simple model of a piece-wise constant stiffness can predict major features of the running exercise makes it an effective tool for future designing of artificial legs and robots and also for the development of more accurate control strategies.

Author details

J. Mizrahi and D. Daily
Department of Biomedical Engineering, Technion, Israel Institute of Technology, Haifa, Israel

Acknowledgement

This Chapter is partly based on an MSc Thesis of second author DD, carried out in JM's Biomechatronics Laboratory, Department of Biomedical Engineering, Technion – Israel Institute of Technology, under the joint supervision of JM and Prof. Yacov Ben-Haim.

8. References

[1] Beck B R. Tibial Stress Injuries. An Aetiological Review for The Purposes of Guiding Management. Sports Medicine 1998;26: 265-279.

[2] Burr DB. Bone, Exercise and Stress Fracture. In: Holloszy JO. (ed.) Exercise and Sport Sciences Review. Baltimore (MD): Williams and Wilkins; 1997, p. 171-194.

[3] Yoshikawa T, Mori S, Santiesteban AJ, Sun TC, Hafstad E, Chen J, Burr DB. The Effects of Muscle Fatigue on Bone Strain. *J. Exp.Biol.* 1994;188: 217-233.

[4] Reeder M T, Dick B H, Atkins JK, Probis AB, Martinez JM. Stress Fractures: Current Concepts of Diagnosis and Treatment. Sports Med 1996;22: 198-212.

[5] Mizrahi J, Voloshin A, Russek D, Verbitsky O, Isakov E. The Influence of Fatigue on EMG and Impact Acceleration in Running. Basic Appl. Myol. 1997;7: 111-118.

[6] Mizrahi J, Verbitsky O, Isakov E. Shock Accelerations and Attenuation in Downhill and Level Running. Clinical Biomech 2000;15: 15-20.

[7] Voloshin A, Mizrahi J, Verbitsky O, Isakov E. Dynamic Loading on the Human Musculoskeletal System- Effect of Fatigue. *Clin. Biomechanics* 1998;13: 515-520.

[8] Verbitsky O, Mizrahi J, Voloshin A, Treiger J, Isakov E. Shock Absorption and Fatigue in Human Running. *J. Appl. Biomechanics,* 1998;14: 300-311.

[9] Baker J, Frankel VH, Burstein A. Fatigue Fractures: Biomechanical Considerations. The Journal of Bone and Joint Surgery 1972;54A: 1345-1346.

[10] Nordin M, Frankel V. Biomechanics of Bone. In: Nordin M., Frankel V. (eds). Basic Biomechanics of the Musculoskeletal System. Philadelphia (PA): Lea and Febiger, 1989, p. 3-29.

[11] Mizrahi J, Verbitsky O, Isakov E. Fatigue-Related Loading Imbalance on the Shank in Running: A Possible Factor in Stress Fractures. *Annals Biomed. Eng.,* 2000;28: 463-469.

[12] Greene PR, McMahon TA. Reflex Stiffness of Man's Anti-Gravity Muscles During Kneebends while Carrying Extra Weights. *J. Biomechanics,* 1979;12: 881-891.

[13] Mizrahi J, Susak Z. In-Vivo Elastic and Damping Response of the Human Leg to Impact Forces. *J . Biomech. Engng.* 1982;104: 63-66.

[14] Ozguven HN, Berme N. An Experimental and Analytical Study of Impact Forces during Human Jumping. *J Biomechanics* 1988;21: 1061-1066.

[15] Kim W, Voloshin AS, Johnson SH. Modeling of Heel Strike Transients during Running. *Human Movement Science,* 1994;13: 221-244.

[16] Farley CT, Gonzalez O. Leg Stiffness and Stride Frequency in Human Running. *J. Biomechanics,* 1996;29: 181-186.

[17] Farley C T, Morgenroth D C. Leg Stiffness Primarily Depends on Ankle Stiffness during Human Hopping. Journal of Biomechanics 1999;32: 267-273.

[18] Spagele T, Kistner A, Gollhofer A. Modeling, Simulation and Optimization of a Human Vertical Jump. ASME Journal of Biomechanical Engineering 1999;32: 521-530.

[19] McMahon TA, Green PR. The Influence of Track Compliance on Running. *J Biomechanics,* 1979;12: 893-904.

[20] Rapoport S, Mizrahi J, Kimmel E, Verbitsky O, Isakov E. Constant and Variable Impedance of the Leg Joints in Human Hopping. *ASME Journal of Biomechanical Engineering* 2003;125: 507-514.

[21] Cavanagh PR, Valiant GA, Misevich KW. Biological Aspects of Modeling Shoe/Foot Interaction during Running. In: Frederick EC. (ed.) Sport Shoes and Playing Surfaces. Champaign: Human Kinetics; 1984. p24-46.

[22] McMahon TA, Valiant G, Frederick EC. Groucho Running. *J Appl. Physiol.* 1987;62: 2326-2337.

[23] Nigg BM. Experimental Techniques Used in Running Shoe Research. In: Nigg BM. (ed.). Biomechanics of Running Shoes. Champaign: Human Kinetics Publishers; 1986; p27-62.

[24] Winter DA. Biomechanics and Motor Control of Human Movement, 2nd Ed., John Wiley & Sons; 1990. p51-74.

[25] Dickinson JA, Cook SD, Leinhardt TM. The Measurement of Shock Waves Following Heel Strike while Running. *J. Biomech.* 1985;18: 415-422.

[26] Clarke T, Cooper L, Clark D, Hamill C. The Effect of Varied Stride Rate and Length upon Shank Deceleration during Ground Contact in Running. *Med. Sc. Sports Exerc.,* 1983;15, 170.

[27] Valiant GA. Transmission and Attenuation of Heelstrike Accelerations. In: P.R Cavanagh (Ed.), Biomechanics of Distance Running . Champaign, IL: Human Kinetics; 1990. p 225-247.

[28] Hamill J, Derrik TR, Holt KG. Shock Attenuation and Stride Frequency during Running. *Human Movement Science,* 1995;14: 45-60.

[29] Lafortune MA, Henning E, Lake MJ. Dominant Role of Interface over Knee Angle for Cushioning Impact Loading and Regulation Initial Leg Stifness. *J Biomechanics,* 1996;29: 1523-1529.

[30] Streitman A, Pugh J. The Response of the Lower Extremity to Impact Forces. I. Design of an Economical Low Frequency Recording System for Physiologic Waveforms. Bulletin of the Hospital for Joint Diseases 1978; XXXIX(1): 63-73.

[31] Wasserman K. Determinants and Detection of Anaerobic Threshold and Consequences of Exercise above It. Circulation 1987; 76(suppl VI), VI-29.

[32] Edwards R H. Human Muscle Function and Fatigue. Human Muscle Fatigue: Physiological Mechanisms. London: Pitman Medical, (Ciba Foundation symposium 82) 1981, p 1-18.

[33] Wright TM, Hayes WC. Tensile Testing of Bone Over a Wide Range of Strain Rates: Effects of Strain Rate, Micro-Structure and Density" Medical and Biological Engineering and Computing 1980;14a: 671-680.

[34] Peterson R H, Gomez MA, Woo S L-Y. The Effects of Strain Rate on the Biomechanical Properties of the Medial Collateral Ligament: A Study of Immature and Mature Rabbits. Transactions of the Orthopedic Research Society 1987;12: p 127.

[35] Li JT, Armstrong CG, Mow VC. The Effects of Strain Rate on Mechanical Properties of Articular Cartilage in Tension, Proc Biomechanical Symposium ASME AMD 1983;56: p 117-120.

[36] Herzog W, Leonard TR. Validation of Optimization Models that Estimate the Forces Exerted by Synergistic Muscles. Journal of Biomechanics 1991;24(S1): 31-39.

[37] Mizrahi J, Ramot Y, Susak Z. The Dynamics of the Subtalar Joint in Sudden Inversion of the Foot, *J Biomech. Engng,* 1990;112: 9-14.

[38] Marquardt DW. An Algorithm for Least Squares Estimation of Nonlinear Parameters. SIAM J.1963;11: 431-441.

[39] Bard Y. *Nonlinear Parameter Estimation.* Academic Press, Inc.; 1974 p 83-217.

[40] Alexander RMcN. The Spring in Your Step: the Role of Elastic Mechanism in Human Running. Amsterdam: Free University press, 1988 p 17-25.

[41] Blickhan R. Full RJ. Locomotion Energetics of the Ghost Crab. II. Mechanics of the Centre of Mass during Walking and Running", Journal of Experimental Biology 1987;130: 155-174.

[42] Blickhan R. The Spring-Mass Model for Running and Hopping. Journal of Biomechanics 1989;22: 1217-1227.

[43] Cavagna GA, Sailbene FP, Margaria R. Mechanical Work in Running. Journal of Applied Physiology 1964;19: 249-256.

[44] Cavagna GA, Heglund NC, Taylor CR. Mechanical Work in Terrestrial Locomotion: Two Basic Mechanisms for Minimizing Energy Expenditure. American Journal of Physiology 1977; 233: R243-R261.

[45] McMahon TA, Cheng GC. The Mechanics of Running: How Does Stiffness Couple with Speed? *J Biomechanics,* 1990;23: 65-78.

[46] Thys H. Evaluations Indirecte de l'Energie Elastique Utilisee dans l'Impulsion des Sauts. Schweizerischen Zeitschrift fur Sportsmedizin, 1978;4: 169-177.

[47] Bosco C, Komi PV.Mechanical Characteristics and Fiber Composition of Human Leg Extensor Muscles. European Journal of Applied Physiology 1979;41: 275-284.

[48] De Clercq D, Aerts P, Kunnen M. The Mechanical Characteristics of the Human Heel Pad during Foot Strike in Running: An In-Vivo Cineradiographic Study. *J. Biomechanics,* 1994;27: 1213-1222.

[49] Bennett MB, Ker RF. The Mechanical Properties of the Human Subcalcaneal Fat Pad in Compression", *J Anat.,* 1990;171: 131-138.

[50] Aerts P, De Clercq D. Deformation Characteristics of the Heel Region of the Shod Foot During a Simulated Heel Strike: The Effect of Varying Midsole Hardness", *J Sports Set,* 1993;11: 449-461.

[51] Kinoshita H, Ogawa T, Kuzuhara K, Ikuta K. In vivo examination of the dynamic properties of the human heel pad. *Int. J Sports Med.,* 1993;14: 312-319.

[52] Aerts P, Ker RF, De Clercq D, Ilsley DW, Alexander RMcN. The Mechanical Properties of the Human Heel Pad: A Paradox Resolved. *J Biomechanics,* 1995;28: 1299-1308.

[53] Alexander RMcN, Bennet MB, Ker RF. Mechanical Properties and Function of the Paw Pads of Some Mammals. *J Zoo1.,* 1986;A209: 405-419.

[54] Lafortune MA, LakeMJ, Henning E. Differential Shock Transmission Response of the Human Body to Impact Severity and Lower Limb Posture. *J Biomechanics,* 1996;29: 1531-1537.

[55] He J, Kram R, McMahon TA. Mechanics of Running under Simulated Low Gravity. *J. Appl. Physiol.,* 1991;71: 863-870.

[56] Jorgensen U, Ekstrand J. Significance of Heel Pad Confinement for the Shock Absorption at Heel Strike. *Int. J Sports Med.,* 1988;9: 468-473.

[57] Ward-Smith AJ. The Bioenergetics of Optimal Performances in Middle-Distance and Long-Distance Track Running. J Biomech, 1999;32: 461-465.

[58] Wasserman K, Whipp BJ, Koyal SN, Beaver WL. Anaerobic Threshold and Respiratory Gas Exchange during Exercise. *J. Appl. Physiol.,* 1973;35: 239-243.

[59] Nielsen J, Bisgård C, Arendt-Nielsen L, Jensen TS. Quantification of Cerebellar Ataxia in Movements of the Hand. In: *Biomechanics*, Seminar 8, Göteburg, Sweden, 1994; p157-166.

[60] Weiss PL, Hunter IW, Kearney RE. Human Ankle Joint Stiffness Over the Full Range of Muscle Activation Levels. *J. Biomechanics*, 1988;21: 539-544.

Section 4

Quantitative Biomechanics

Chapter 10

Correlating Micro-CT Imaging with Quantitative Histology

Tomáš Gregor, Petra Kochová, Lada Eberlová, Lukáš Nedorost, Eva Prosecká, Václav Liška, Hynek Mírka, David Kachlík, Ivan Pirner, Petr Zimmermann, Anna Králíčková, Milena Králíčková and Zbyněk Tonar

Additional information is available at the end of the chapter

http://dx.doi.org/10.5772/48680

1. Introduction

Advanced biomechanical models of biological tissues should be based on statistical morphometry of tissue architecture. A quantitative description of the microscopic properties of real tissue samples is an advantage when devising computer models that are statistically similar to biological tissues in physiological or pathological conditions. The recent development of X-ray microtomography (micro-CT) has introduced resolution similar to that of routine histology. The aim of this chapter is to review and discuss both automatic image processing and interactive, unbiased stereological tools available for micro-CT scans and histological micrographs. We will demonstrate the practical usability of micro-CT in two different types of three-dimensional (3-D) *ex vivo* samples: (i) bone scaffolds used in tissue engineering and (ii) microvascular corrosion casts.

2. Principles of micro-CT

This chapter covers the basic principles of micro-CT. *Ex vivo* specimens are typically placed on a rotating stage between the X-ray source and the microscope objective, which is followed by a detector (Fig. 1). For high resolution imaging, the sample size must be reduced to a minimum. The dimensions should not exceed 500-1000 times the resolution limit required. In large samples, the X-rays must penetrate more material, which results in a lower photon count and increased exposure time.

Certain devices operate with geometrical magnification only, in which the resolution increases with the distance between the sample and the detector. Unfortunately, increased geometrical magnification can result in blurriness, depending on the X-ray source spot size.

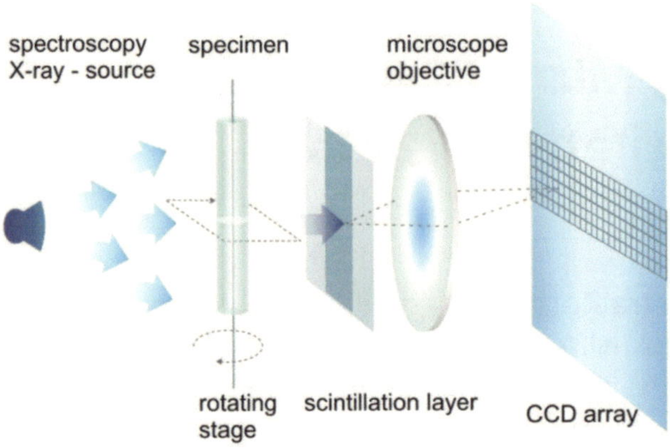

Figure 1. Configuration of a micro-CT scanner with an *ex vivo* sample rotating within a stationary X-ray system (redrawn and modified according to Jorgensen et al., 1998).

More sophisticated devices use microscope objectives for increased magnification and resolution. For optimal settings of the micro-CT scan, the spot size (SS) and the awaited pixel size (PS) are the parameters that set the source distance (SD) and detector distance (DD). When the SD and DD are smaller, higher photon counts can be achieved, thus reducing the time costs. The settings should fulfill equation 1; see also Fig. 2:

$$PS \geq SS \cdot \frac{DD}{SD} \tag{1}$$

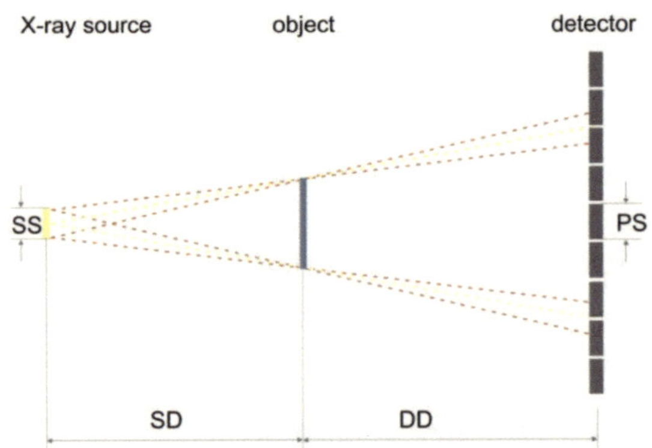

Figure 2. For optimal settings of the micro-CT scan, the spot size (SS) and the awaited pixel size (PS) are the parameters that set the source distance (SD) and detector distance (DD) (redrawn and modified according to Roth et al., 2010).

The source SS is strongly dependent on the power of the X-ray tube used. In today's machines, the source SS ranges between 1-10 µm. For high resolution scanning of biological samples, it is advisable to operate with a low accelerating voltage (the typical range for the Xradia XCT 400 (Xradia, Pleasanton, CA, USA) is 20–60 kV used with a power of 4 W, at which the tube yields the lowest SS). The scanning time depends on the magnification and resolution required. For details smaller than 1 μm, 24 hours or more might be necessary, whereas an overview scan with a pixel size of approximately 10 μm can be achieved within an hour. Sample drifting might be an issue during a long scanning time. From the reconstructed images, the objects of interest are visualized, thresholded, traced and analyzed.

3. Current applications of micro-CT in biomechanics and medicine

Compared with standard human CT devices, which offer a resolution limit of approximately 0.4 mm, the micro-CT introduced a promising modality. However, the clinical use of this method is limited by its higher radiation exposure and longer scanning times. It is used either to visualize individual fine functional and anatomical structures of *ex vivo* human or animal organs (e.g., liver lobules or bone trabeculae) or for whole-body imaging of small animals (Schambach et al., 2010). *In vivo* micro-CT systems are based on a rotating system of X-ray tube and detectors. The construction of these devices is the same as in human CT, except that their dimensions are adapted to small animals (Bag, 2010). The minimal space resolution of *in vivo* micro-CT is from 100 to 30 μm. The imaging of living animals must be faster than in *ex vivo* micro-CT. It is enabled by, among other factors, the use of flat panel detectors that allow us to acquire an abundance of thin sections during one rotation. The examination is also limited by the necessity of using a radiation dose that does not harm the tested animal. In *ex vivo* micro-CT, the X-ray source and the detectors are stationary, and it is possible to adjust the distance between the X-ray tube and the detectors (based on the size of the examined object) to improve the spatial resolution and minimize artifacts. Moreover, in *ex-vivo* micro-CT, time resolution is not important; thus, the examination may take a very long time (hours), and any amount of radiation may be used. This technique enables us to acquire much higher spatial resolution than *in vivo* micro-CT (30 to 1 μm) (Zagorchev et al., 2010).

Most organs have already been analyzed with micro-CT, including bones (Peyrin, 2011), heart and blood vessels (Schambach et al., 2010), lungs, kidney, liver, and cerebral structures (Schambach et al., 2010). Micro-CT devices can be used for the characterization of bone or vascular microarchitecture (Peyrin, 2011; Burghardt et al., 2011; Missbach-Guentner et al., 2011). This method also allows the precise detection of the margins of tumors and their vascularity (Ma et al., 2011; Missbach-Guentner et al., 2011). Tissue composition (e.g., bone mineralization) can be directly linked to 3-D tissue morphometry (Burghardt et al., 2011). Thus, 3-D micro-CT analysis becomes a method of choice for describing the spatial complexity of organ segmentation and the relationships between morphological and functional units (e.g., hepatic lobules and portal acini) (Schladitz, 2011). Micro-CT can also link the imaging of anatomical structures with functional and molecular imaging, e.g., tissue and organ perfusion, the flow rate of exocrine secretions within parenchymatous organs and glands (Marsen et al., 2006), or heart movements (Badea et al., 2005).

For the examination of soft tissues and vessels, contrast solutions are necessary. In *in vivo* imaging, such as in human medicine, standard iodinated contrast media or intravascular blood-pool contrast agents are used (e.g., contrast material covered by a polyethylene glycol capsule and stabilized by lipoproteins or iodinated triacylglycerides). These blood-pool contrast media are able to remain in the blood circulation for a longer time and thus enable longer scanning times. They do not leave the blood circulation as do standard iodinated contrast agents, which pass into the extravascular interstitial space. Contrast substances labeled with antigens or other ligands (Ritman, 2011) may also be used for targeting and tracking specific structures, such as stem cells (Villa et al., 2010). For *in vivo* studies, nanoparticles can be used to enhance the soft tissue contrast (Boll et al., 2011). It is also possible to use nanoparticles that incorporate into, e.g., tumors and could remain there for longer periods (Boll et al., 2011). However, *ex vivo* micro-CT can utilize any contrast solution, including those that are toxic to living organisms. The only limitation is that they must not damage the examined tissue. In *ex vivo* micro-CT imaging, it is recommended that contrast solutions be used that offer high contrast to the studied tissue, that have a low viscosity to fill the smallest vessels and that do not diffuse out of the blood vessels. In practice, the substances used include silicon rubber (Savai et al., 2009), polymethylmethacrylate with added lead pigment, and gelatin with bismuth or barium sulfate (Zagorchev et al., 2010).

Other modifications of CT applicable in medical experiments are the mini-CT devices (voxel size 10^{-3} mm^3, used for scans of whole organs or small animals) and the nano-CT devices (voxel size of 10^{-7} mm^3) (Ritman, 2011; Müller, 2009).

The greatest progress in micro-CT exploration was acquired in the exploration of tumor microvascularization and the study of neoangiogenesis. The latter phenomenon is also important for the understanding of tumor growth and could be used in oncological treatment strategies, especially in patients treated by biological therapy with antibodies against vascular endothelial factor A (bevacizumab), which inhibits neoangiogenesis (Ma, 2011). Micro-CT imaging of pathological vascularity can provide new information, e.g., about changes in vessel walls in atherosclerosis or other pulmonary vascular diseases (Razavi, 2012).

A promising trend in experimental work is using hybrid methods that combine detailed anatomical information from micro-CT with information about cellular metabolism and structure from methods of nuclear medicine (micro-SPECT/CT and micro-PET/CT) (Ritman, 2011). Therefore, it is appropriate to combine or compare the results of micro-SPECT or micro-PET with, e.g., microscopic analysis of a specimen to estimate the correct anatomical orientation and acquire a satisfactory interpretation of the results. The new suggested tools would be able to use higher energy examination from more X ray sources and thus obtain results on the cellular or subcellular level. The development of new tissue-specific contrast solutions could also be promising for future research activities using micro-CT or its hybrid methods.

4. Micro-CT imaging of biological samples *ex vivo*

4.1. Biomechanics of bone scaffolds

The aim of this chapter is to illustrate the application of micro-CT in tissue engineering and in assessing the biomechanics of biocompatible collagen/hydroxyapatite bone scaffold samples (Prosecká et al., 2011). Tissue engineering is a promising interdisciplinary research field that aims to develop biological substitutes for the repair of damaged tissue. The typical strategy involves either the delivery of isolated and expanded cell populations within a tissue engineering construct or the recruitment of host cells local to the site of damage through the use of conductive scaffolds and inductive biological signals. The question of how to optimize the design of scaffolds for different tissues remains unsolved. To assess the suitability of polymer tissue scaffolds for use in regenerative medicine, methods to characterize scaffolds are needed (Renghini et al., 2009). The bone scaffolds should be stiff enough to withstand high forces in the bone after implantation but, conversely, should also be flexible enough to enable growth of the cells and changing of the original shape of the graft to meet the needs of the bone complex. Therefore, it is important to perform mechanical measurements and 3-D imaging of bone scaffolds before the seeding of mesenchymal stem cells.

Generally, bones and bone scaffolds can be mechanically tested using various types of techniques: tensile or torsion tests used for strip- or block-shaped tissue specimens; a pressure test used for block-shaped or cylindrical specimens; a ring test in which a ring of given thickness is cut from a tubular organ (typically a blood vessel), clamped into the jaws of special measurement devices and loaded by tension; and an intraluminal pressure inflation-deflation test of tubular organs. The choice depends on the physiological loading of the tissues. The aim is to be as close as possible to the real loading and thus to the real mechanical properties. In any case, regardless of the chosen technique, the result given by a measurement device is of the same nature: a stress-strain (tensile, pressure, ring test) or pressure-outer diameter (inflation-deflation test) curve. The stress-strain curve has a mostly nonlinear shape showing the tissue stiffening as loading increases. The stiffening is caused by various tissue components as they subsequently contribute to the tissue response. The soft component, mostly elastin, contributes to the mechanical response at low loading and is connected with the low stiffness of the tissue, whereas the curly and stiff collagen fibers are straightened as loading progresses and contribute to stiffening at high loading. The relevant portions of a stress-strain curve could be approximated (e.g., by a line), and thus, the mechanical parameters, such as Young's modulus of elasticity at small deformations (low loading) and at large deformations (high loading), the pressure-strain elastic modulus, the initial modulus of compression, the limit stress and the strain in the case of loading until tissue rupture, could be obtained.

To emulate the loading of bones, in which most of their parts are under pressure and only a small fraction of the tissue is under tension, pressure loading was applied to collagen/hydroxyapatite composite bone scaffolds. Cylindrical specimens (approximately 12 mm in diameter and in height) of composite scaffolds containing various amounts of collagen and hydroxyapatite prepared according to Prosecká et al. (2011) underwent pressure mechanical loading with a loading velocity of 1 mm/min. The resultant stress-strain curves, and particularly the regions between 2% and 10% of the original specimen's height, were approximated by linear regression, and thus the initial moduli of compression were determined. (for details of the measurements and their evaluation, see Prosecká et al., 2011).

From a biomechanical point of view, the shape of the stress-strain curves was identical for all compositions of scaffolds. The beginning of each stress-strain curve was linear with low stiffness, and increasing loading led to scaffold stiffening. This stiffening was likely caused by the fact that, with increasing deformation of the porous specimens, the originally high amount of free space filled by air became smaller, and the stiff components became closer together and started to contribute more to the mechanical response. The decreasing porosity caused by the increasing pressure loading could thus be connected with the higher stiffness of the scaffold. The initial modulus of compression increased rapidly with an increase in the collagen concentration. From these two results, we can conclude that not only the composition but also the porosity may play crucial roles in the mechanical properties of collagen/hydroxyapatite composite bone scaffolds.

4.2. Imaging and biological applications of bone scaffolds

The 3-D porous structure and sufficient mechanical stiffness of the bone scaffolds are necessary conditions for the attachment, growth, and progress of mesenchymal osteoprogenitor stem cells (Prosecká et al., 2011); see Fig. 3.

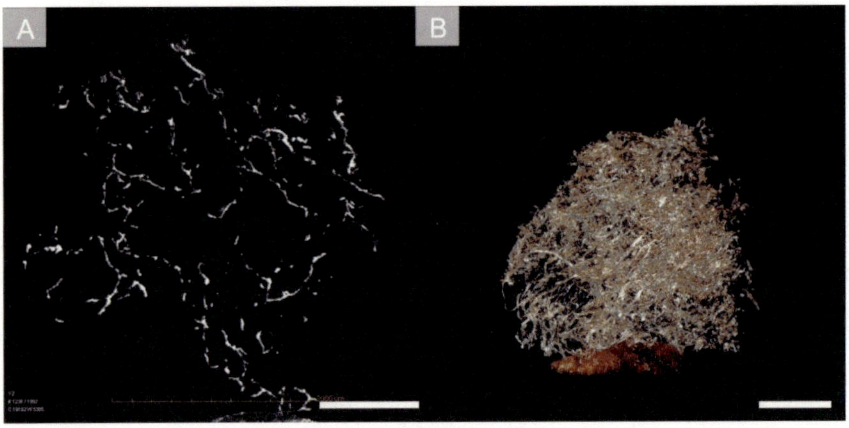

Figure 3. Micro-CT scans of a bone scaffold manufactured from collagen/hydroxyapatite matrix. A – a single section acquired with a 4× objective; the scanning time was 9 hours. B – an image reconstruction based on 3376 sections. The scale bars indicate 1 mm.

Traditional methods for evaluating the osseointegration of tissue-engineered scaffold/cell constructs are based on 2-D histological and radiographic techniques. Sectioning followed by histology can image the scaffold interior but is destructive, lengthy and only semi-quantitative (Ho & Hutmacher, 2006). Fluorescence microscopy can be quantitative when high-throughput approaches are applied, e.g., producing 3-D images with confocal fluorescence microscopy (Tjia & Moghe, 1998). Colorimetric and fluorometric soluble assays are available for cell components, such as enzymes, protein or DNA (Ho & Hutmacher, 2006). However, these soluble assays are quantitative but do not provide information on cell distribution. In contrast to these methods, by micro-CT we can non-destructively obtain 3-D images that penetrate deep into the scaffold interior and produce inherently quantitative results (Tjia & Moghe, 1998). Micro-CT currently appears to be the most suitable approach for this task (Ho & Hutmacher, 2006; Mather et al., 2007; Mather et al., 2008; Cancedda et al., 2007). Clearly, in a structure as complex as bone, micro-CT provides a distinct advantage over conventional microscopy. Structures can be followed continuously, from the level of osteon to gross bone morphology.

Dorsey et al. (2009) explored the use of X-ray microcomputed tomography for observing cell adhesion and proliferation in polymer scaffolds. The ability of micro-CT to detect cells in scaffolds was compared with those of fluorescence microscopy and a soluble DNA assay. The researchers demonstrated that fluorescence microscopy had better resolution than micro-CT and that the soluble DNA assay was approximately 5 times more sensitive than micro-CT. However, by micro-CT, they were able to reveal the interiors of opaque scaffolds and to obtain quantitative 3-D imaging and analysis via a single, non-invasive modality. They observed that for quantitative micro-CT volume analysis, a cell density of more than 1 million cells/ml is required in the scaffolds. The results demonstrate the benefits and limitations of using micro-CT for the 3-D imaging and analysis of cell adhesion and proliferation in polymer scaffolds. Various bone engineering groups have noted the importance of micro-CT analyses in tissue engineering. Among them, Müller and Rüegsegger have investigated, and quantified, the architecture of cancellous bone using micro-CT (Müller, 1994; Müller et al., 1996, 1998; Rüegsegger, 2001).

In addition to these analyses, a number of morphological measures are being investigated, such as volume fractions of tissues, directivity of calcifying tissue, porosity, pore connectivity, putative vascularization, curvature and surface-to-volume measure (Bentley et al., 2002; Jorgensen et al., 1998).

Despite the advantages of using micro-CT, there are still several issues with both image segmentation and resolution that are exacerbated by the low image contrast due to the low X-ray attenuation of the materials being used (Mather et al., 2008). Morris et al. (2009) studied a method for the generation of computer-simulated scaffolds that resemble foamed polymeric tissue scaffolds. They showed that the quality of the images (and hence the accuracy of any parameters derived from them) may be improved by a combination of pixel binning and by taking multiple images at each angle of rotation. However, micro-CT is considered to be a standard technique in tissue-engineered bones (Cancedda et al., 2007).

5. Three-dimensional imaging of microvascular corrosion casts

5.1. Corrosion casting

Anatomical corrosion casts provide 3-D insights into the arrangement of hollow structures and organ cavities at both the macro- and microscopic levels. The most frequently used are vascular corrosion casts (Giuvărăşteanu, 2007), which, in combination with scanning electron microscopy, constitute the primary application of corrosion casting describing the morphology and anatomical distribution of blood vessels (Lametschwandtner et al., 2004; 2005). Many other examples are found in the literature - the airways (Hojo, 1993), bile ducts

(Gadžijev & Ravnik, 1996), urinary tract (Marques-Sampaio et al., 2007) and lymphatic system (Fujisaka et al., 1996) have been visualized by means of corrosion casting. The steps for obtaining the corrosion casts are as follows: perfusion of the hollow target structures to remove the contents (in the case of vascular casts, prior heparinization is required for the maintenance of blood fluidity), injection of the casting media, its polymerization in the fully filled cavities and the subsequent removal of the surrounding tissues by a highly aggressive corrosive solution. The corrosion casting method has been known since the beginning of the 16th century, when Leonardo da Vinci injected dissolved wax into bovine cerebral chambers (Paluzzi et al., 2007). Over the following centuries, all of the steps in the method have been improved. However, the aim has remained the same – to obtain the most accurate replica of the biological structure.

5.2. Casting media

Because the highest authenticity is required in the proportions of 3-D corrosion casts, casting media must have sufficient viscosity (to pass through but not to penetrate); they must be capable of even, rapid polymerization with minimal shrinkage and of physicochemical resistance to the subsequent corrosion and dissection procedures. All of these properties are combined in methylmethacrylate (MMA) resin. The MMA polymerizing fluid can be produced by warming the MMA monomer with a catalyst, benzoyl peroxide. MMA easily penetrates the capillary network and flows out via the opposite vessels (Kachlík & Hoch, 2008). A polyurethane pigment paste of various colors can be added into the mass along with the X-ray contrast medium, e.g., barium sulfate ($BaSO_4$) or mercury sulfide (HgS) (Gadžijev & Ravnik, 1996). There are also commercially produced, partially polymerized MMAs (e.g., Mercox or Dentacryl), but these products vary considerably in their rheological properties. Dentacryl was used mainly in the last century, originally in dental prosthetics. This compound is a commercially produced, colorless, partially polymerized MMA that is not, due to its high viscosity, suitable for casts of the microvasculature (Kachlík & Hoch, 2008). The greatest advantage of this casting medium is its low cost; thus, Dentacryl is used for more spacious casts, such as those of the bronchial tree (Havránková et al., 1989) or paranasal sinuses (Hajnis, 1988). Painting clays can be used to color Dentacryl. Mercox is a low-viscosity acrylic resin, a MMA monomer stabilized solution. Mercox II (Ladd Research, Williston, Vermont, USA) is commercially offered in two colors (blue and red); the kit contains an MMA resin and a catalyst (benzoyl peroxide). The Mercox II system combines excellent permeability through the entire vascular bed, excellent infiltration properties, short time of preparation and polymerization, minimal shrinkage and high chemical resistance. Although its price is rather high, it is a method of choice for the vascular corrosion casts (Bartel & Lametschwandtner, 2000; Minnich et al., 2002; Lametschwandtner et al., 2004; Kachlik & Baca, 2006, Kachlík & Hoch, 2008).

In our experience with obtaining corrosion casts of porcine liver vascular trees, it is helpful to administer 50,000 IU of heparin in 1 liter of saline and subsequently rinse the vascular bed with 5 liters of Ringer´s solution prior to sacrificing the animal. We found that even the lowest Mercox dilution recommended by the manufacturer (20 ml resin/0.4 g of 40%

benzoylperoxide) polymerizes within a very short time (approximately 5 min), which makes it difficult to fill the complicated liver vasculature (Eberlová, 2012, unpublished results). After 5 minutes of stirring the resin with the catalyst, it was possible to fill a volume of approximately 20 ml before the Mercox started to harden. To prevent any vascular lesions and artifacts, the use of an infusion pump appears to be warranted (Minnich et al., 2002).

Unfortunately, Mercox does not offer sufficient X-ray-opacity for micro-CT. To fill and opacify microvessels *ex vivo*, the silicone rubber Microfil MV (Flow Tech, Inc., Carver, Massachusetts) appears to be a substance of choice (Gössl et al., 2003). The Microfil MV kit is available in five radio-opaque colors and comprises MV-Diluent, MV compound and MV Curing Agent. The working time of Microfil is approximately 20 minutes. However, Microfil MV is not suitable for conventional corrosion casting techniques using potassium hydroxide. Two alternative techniques for the subsequent tissue clearing are offered instead: alcohol-methyl salicylate clearing, which produces a stiffer tissue useful for gross observation, and glycerin clearing, which yields a more flexible tissue and vessels.

6. Automatic image processing, topology analysis and measurement of statistical features

This chapter describes an approach to automatic micro-CT image processing using computer vision techniques and the Mercox vascular corrosion casts of the intestinal mucosa described in the previous chapter. The input data were acquired with Xradia XCT 400 (Xradia, Pleasanton, CA, USA) by scanning Mercox vascular corrosion casts. The resulting data set consisted of a DICOM stack of approximately 1000 slices, each 1000 times 1000 pixels, such that the scanned volume contained 10^9 volume elements (voxels), each represented by a 16-bit signed number. The depicted volume in reality was a cube with an edge of approximately 35 mm, whereas the voxel edge was 36.53 μm. For image processing, we used the Insight Toolkit library for C++ (www.itk.org) and MATLAB with its Image Processing Toolbox (The MathWorks, Inc., Natick, MA, USA). The first step of the image processing was an analysis of the density distribution of the data, using a density histogram; see Fig. 4.

Segmentation is a process in computer vision that divides the voxels of the source image into two subsets – foreground and background – depending on their affinity to the objects of the real world (Sonka et al., 1998). In our case, the foreground was formed by the vascular corrosion cast, whereas all of the other data were considered to be background.

Thresholding is a common and widely used image segmentation method in machine vision. This technique outputs a binary image *g(i,j,k)* that classifies the voxels of the source image *f(i,j,k)*, in which *i, j, k* represent the spatial indices of the data, according to equation 2:

$$\begin{aligned} g(i,j,k) &= 1 \quad for\ f(i,j,k) \geq T \\ &= 0 \quad for\ f(i,j,k) < T \end{aligned} \quad . \tag{2}$$

In this equation, T is the threshold value; $g(i,j,k)=1$ for image elements of objects and $g(i,j,k)=0$ for image elements of the background (Sonka et al., 1998). The threshold value *32,768*, used for segmentation in our case, is highlighted in red in Fig 4.

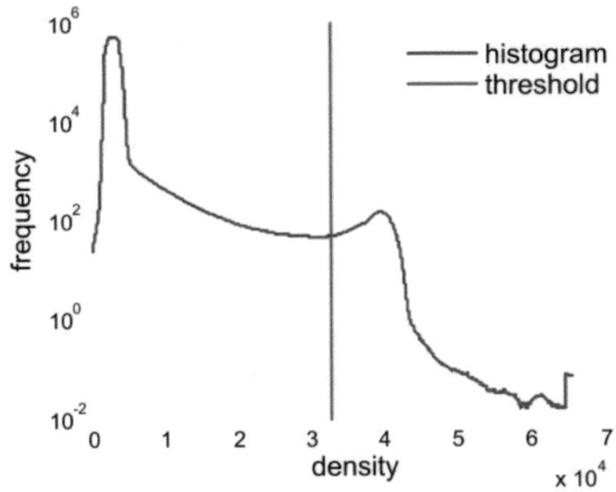

Figure 4. A density histogram of the source image data. In this histogram, which has been smoothened for visualization, we can observe a local minimum prior to a peak representing the density of desired vessel voxels. According to the histogram, the segmentation methods based on density values appear to be suitable for the current image data set.

After segmentation was performed, it was necessary to further pre-process the binary image to eliminate artifacts, e.g., holes and bays caused by irregular spreading of the casting material, and measurement errors. These errors were resolved by applying the morphological operation of hole filling, which converts cavities inside of objects into object voxels. The next step of the data analysis used the labeling algorithm, which assigned a unique label to each contiguous region of a binary image (Sonka et al., 1998). This procedure gave us a large amount of useful information about object counts and sizes in the volume. Part of the label image histogram is shown in Fig. 5.

The label histogram (Fig. 5) demonstrates that there are many objects of insignificant size that are produced simply by measurement noise. We can convert these objects into background and eliminate them from further processing. The size of the objects at which the segmentation of the blood vessel regions becomes unreliable, reveals also the resolution limits of the current micro-CT scan. Acquiring a reliable representation of all of the microvessels would therefore require another scan with a different objective and micro-CT settings. We can also deduct the count of dominant objects in the volume. For the purposes of visualization, in the subsequent processing steps, we will only work with the largest object. Fig. 6 demonstrates the difference between a 3-D model of the whole vessel system (Fig. 6A), and the largest continuous object only (Fig. 6B).

The volumetric image (Fig. 6) is useful for visualization. However, it is not desirable for topological investigation. A shape simplification is needed that would preserve the tree connectivity and geometric conditions. In this study, we used the algorithm introduced by Lee et al. (1994). This algorithm is based on parallel 3-D binary image thinning, which is

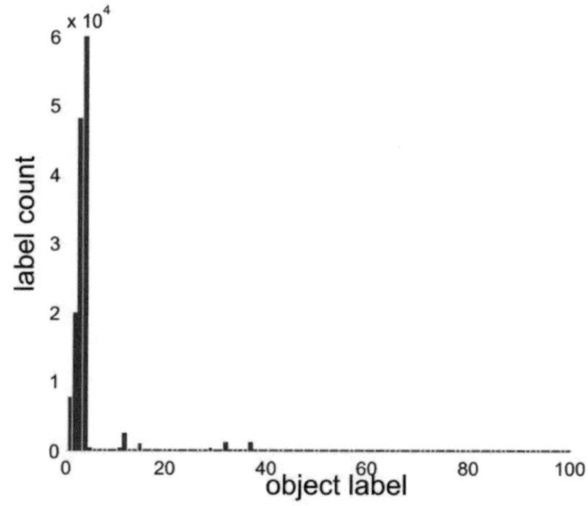

Figure 5. A label image histogram. The object labels each stand for voxels belonging to a single contiguous object, i.e., the count of each label denotes the size of the object it represents, and the number of unique labels corresponds to the number of contiguous objects in the volume. Object labels higher than 100 were omitted for a better overview. There were 1,967 unique objects within the image data set under study.

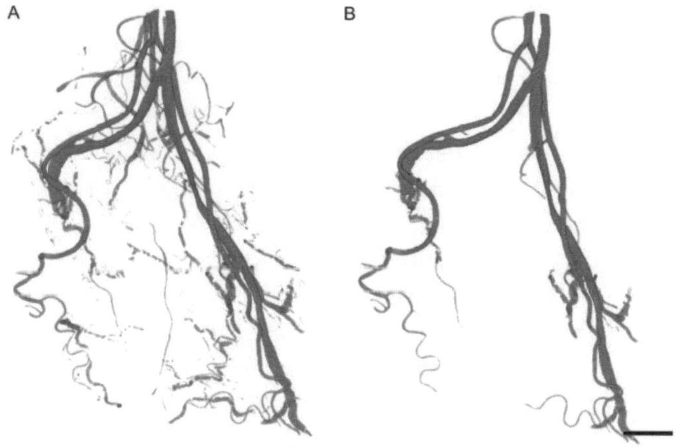

Figure 6. A three-dimensional reconstruction of the entire vessel system (A), and of the largest continuous object (B).

suitable for large data sets and produces a 1-voxel-thin 3-D skeleton. The topology analysis of the skeleton involves identifying the node and terminal points of the tree. The points are located based on their 26-connectivity to neighboring voxels. Based on the properties of the skeletonization algorithm, we can find nodes of valence 3 (i.e., nodes at which three microvascular segments join). Existing nodes of valence 4 are decomposed into two nodes of valence 3. Fig. 7 shows the resulting skeleton with located node points. In this sample, 149 nodes were detected. Using the number of branching points with a known valence, the numerical density of the microvessels *Nv(cap/ref)* can be estimated, as described by Lokkegaard et al. (2001).

Figure 7. Vessel skeleton with branching node points.

Currently, we are able to segment a vessel tree formed by a corrosive cast. In the example analyzed in this study, continuous regions of the volumetric model were counted to detect the number of individual objects within the volume. The volume of each individual object was estimated. Using skeletonization, the number of branching nodes was counted, and the number of vascular segments between the nodes was estimated, as was the length of each segment. Knowing the vessel length and volume, it is also possible to compute the average diameter of the vessel. However, there are still challenges for future work, such as estimating the vessel diameter in each voxel of the skeleton using a distance transform. With this information, the diameter distribution with respect to the vessel length or the vessel volume could be acquired. With a known diameter, the surface area of the vessels can also be computed. Tracking the skeleton leading its binary graph construction determines the spatial distribution of the branching nodes from the proximal vascular segments to the periphery of the vascular tree.

The approach presented in this chapter is based on several assumptions that deserve to be discussed, as they also represent the limitations of this study. We assume, for our computation, that the vessels are in the form of generalized cylinders, which means that the cross-section orthogonal to the medial axis is a circle. The voxels in our data set were cubical; however, a data set with unequal voxel edges may be resampled into cubical voxels.

An important question is the choice of the threshold value. The segmented vessel diameter is partially linearly dependent on the chosen threshold value. Setting an excessively low threshold causes too many artifacts to appear and objects to merge into each other; contrariwise, an excessively high threshold makes vessels very thin, such that small vessels disappear. This effect implies that the threshold value influences the absolute diameter of the vessels and biases statistical markers. However, the construction of the skeleton and the topological analysis do not appear to be affected by the selection of the threshold.

Let us summarize the advantages and disadvantages of our automatic image processing approach. Compared with stereological methods currently in use (see the next chapter), automatic image processing requires virtually no user interaction. This approach is able to compute the distributions of the volumes, surfaces and lengths of vessels of given diameters with a high precision. Moreover, the stereological methods may be applied to our volumetric model to achieve results comparable to those of stereological measurements performed by a human, but with no interaction at all. A disadvantage is the necessity of the choice of the threshold used for the segmentation, due to the sensitivity of certain results to the threshold settings. As stated before, the threshold value does not greatly influence the topology of the vessel tree but strongly correlates with the diameters of the vessels. A proposed solution to this issue is to compute the skeleton with a higher threshold and the diameters with a lower one.

7. Quantitative micro-CT, histology and stereology

Current micro-CT devices are typically bundled with sophisticated software packages that offer a number of automated quantification procedures. However, correlating the micro-CT results with quantitative histology favors the use of unbiased stereological methods, which are highly standardized and widely accepted in biomedical microscopy research (Howard and Reed, 1998; Mouton, 2002). This chapter illustrates the stereological assessment of micro-CT scans of bone scaffolds and microvascular corrosion casts, including the quantification of the volume fraction (V_V, dimensionless), surface density (S_V, m^{-1}), length density (L_V, m^{-2}), orientation and anisotropy of microvessels (Kochová et al., 2011).

In bone tissue samples, the micro-CT resolution is currently capable of providing images that can be used for both analysis of bone vascular canals, and counting individual osteocyte lacunae. Quantification of bone microporosities is used for testing their effect on the viscoelastic properties of bone tissue (Tonar et al., 2011). The microporosity has at least two functional levels, the vascular porosity (related to the vascular canals; the order of magnitude is 10–1000 μm) and the lacunar-canalicular porosity (surrounding the osteocytes; the order of magnitude is 0.1–10 μm). Mechanical experiments clearly demonstrated that the hierarchical organization of bone architecture is crucial and that bone structural heterogeneity varies with the scale of magnification. Whereas Fig. 8 demonstrates two-dimensional sections of compact bone produced by histology and micro-CT, Fig. 9 shows the results of 3-D micro-CT reconstruction of cancellous bone.

The vascular corrosion casts described in section 4 and used in section 5 also can be assessed with spatial stereological methods. Image data acquired by micro-CT are demonstrated in Fig. 10.

After manually tracing the microvessel profiles within a series of consecutive two-dimensional micro-CT sections (software Ellipse, ViDiTo, Košice, Slovak Republic), a three-dimensional system of oriented lines can be acquired (Fig. 11).

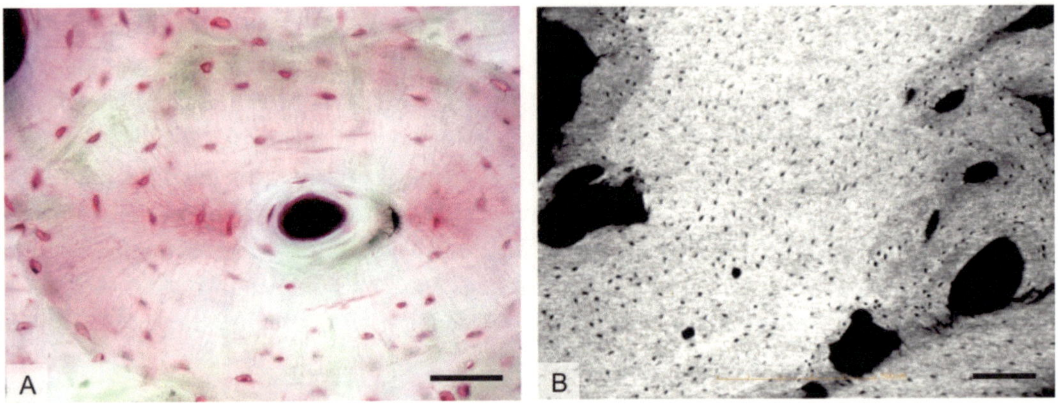

Figure 8. Comparing a histological ground bone section stained with basic fuchsin (A, human femur) with a micro-CT image of compact bone (B, human tibia). In the compact bone, two types of microporosities can be quantified – the osteocyte lacunae and the vascular canals. Both levels of microporosities are clearly visible in either method. The volume fraction of the vascular canals can be quantified stereologically with a point counting method, whereas the numerical density of the osteocyte lacunae can be assessed by the 3-D counting method called disector, which is not biased by the variation in size and orientation of the lacunae (Sterio, 1984; Tonar et al., 2011). The scale bars indicate 60 μm (A) and 200 μm (B).

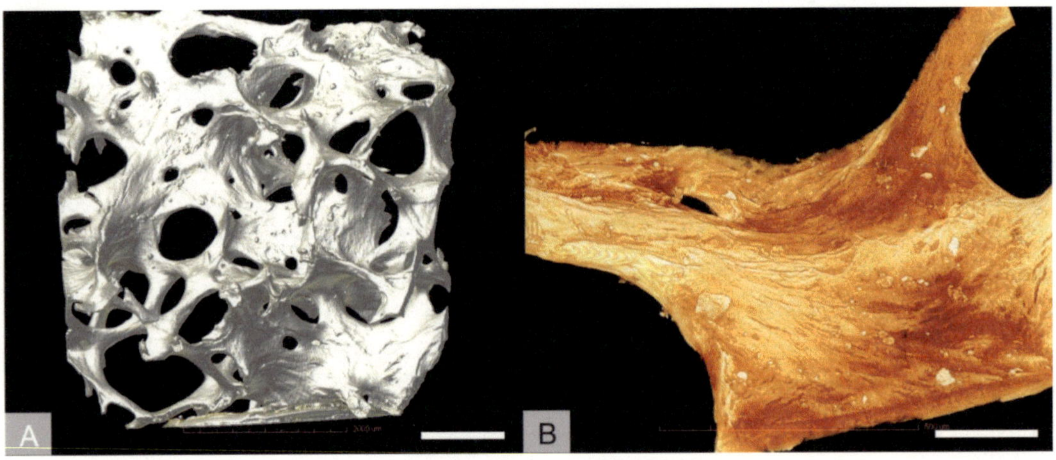

Figure 9. Micro-CT reconstruction of cancellous bone (human tibia) – an overall view (A) and a detail of the surface of bone trabeculae (B). The density and 3-D arrangement of bone trabeculae can be easily assessed with micro-CT. In contrast to scanning electron microscopy of bone surfaces, micro-CT is not biased by perspective or the depth of the 3-D sample. A dry bone sample does not require any laboratory processing prior to micro-CT scanning. The scale bars indicate 1 mm (A) and 200 μm (B).

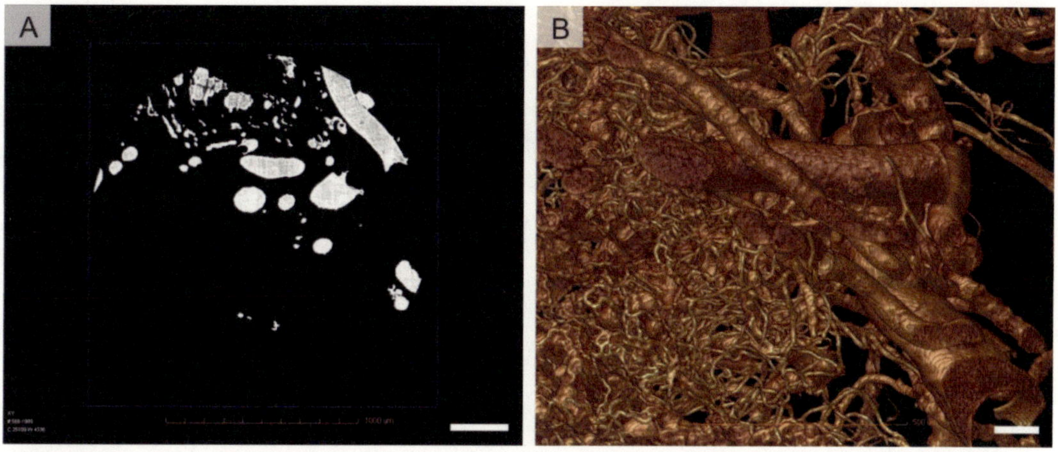

Figure 10. A micro-CT image (A) and a 3-D reconstruction (B) of a vascular corrosion Mercox cast of human intestinal mucosa. The scale bars indicate 300 μm (A) and 100 μm (B).

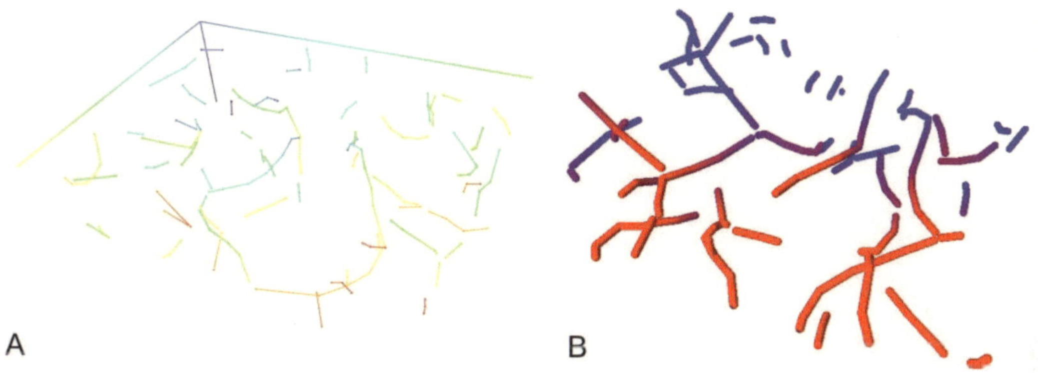

Figure 11. Tracing the microvessel profiles in serial micro-CT sections (Fig. 10A) results in oriented lines, which can be visualized either as linear structures (A) or as rods (B). The blood microvessels are abstracted as having one dimension only (the length), whereas the spatial orientation is retained. The thickness of the rods (B) has been set for better visualization only and does not represent the real thickness of the original microvessels.

Next, the orientation of each skeletonized vessel can be described using a spherical coordinate system (Fig. 12). Each blood vessel segment is described as a vector connecting the center of the coordinate system with the surface of the sphere. This vector is described by its length and a combination of azimuth ranging between [0, 2π] and elevation [0, π/2].

Next, the combinations of vessel lengths and their 3-D orientation can be assessed using various 2-D plots (Fig. 13). These plots are very useful when assessing the directionality and anisotropy of the vascular segments. Should the anisotropy be quantified, several methods are available, such as the ellipsoidal anisotropy, fractional anisotropy, or a chi-square method comparing the observed length densities of lines with the discrete uniform distribution of an isotropic line system (Kochová et al., 2011).

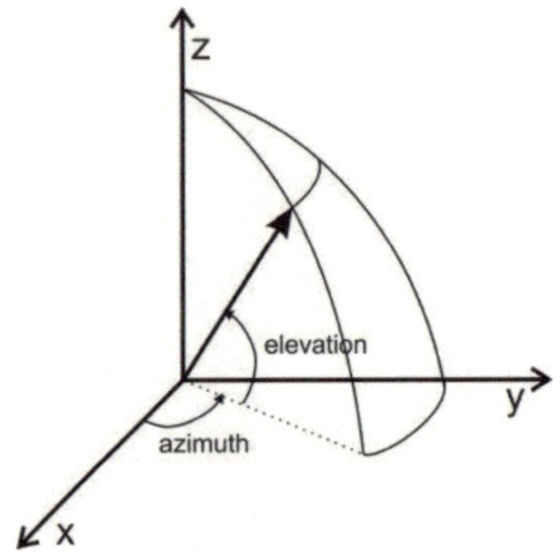

Figure 12. The angular description of the directions of vascular line systems using a spherical coordinate system. Each blood vessel segment is represented by a vector with a known length and a combination of azimuth (longitude) and elevation (latitude). This figure was redrawn and modified according to Kochová et al. (2011).

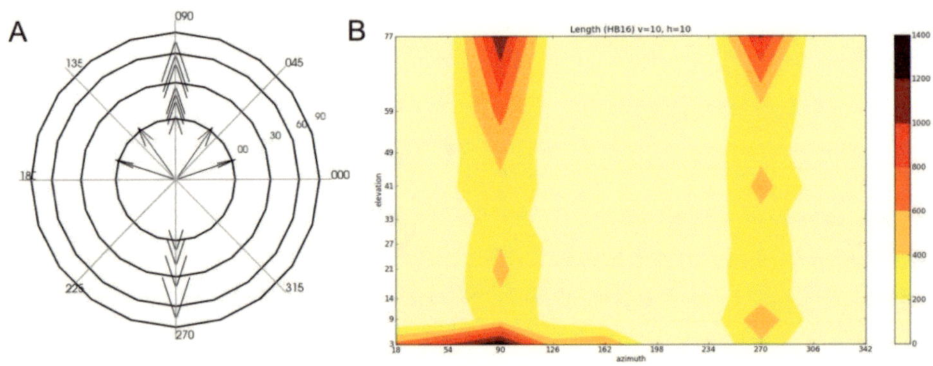

Figure 13. Polar plots can be constructed (A) using the Lambert azimuthal equal area projection. Radial lines are azimuths and concentric lines are elevations, whereas the arrows indicate individual directions. Color histograms can also be used (B) with a color scale corresponding to the microvessel lengths in the given combination of elevation and azimuth (a dark color represents a high value of the microvessel length). In this example, the prevailing directions demonstrate the anisotropy of the microvessels, as both plots exhibit preferential combinations of azimuth and elevation.

Using the skeletons of the microvessels, their lengths L within a reference volume *V(ref)* can be expressed as the length density *Lv*; see equation 3:

$$L_V = \frac{L}{V(ref)} \,. \tag{3}$$

The volume fraction occupied by the microvascular corrosion cast easily can be estimated using the Cavalieri principle (Howard & Reed, 2005), as shown in equation 4,

$$estV = T \cdot (A_1 + A_2 + \ldots + A_m) \,, \tag{4}$$

where $estV$ is the estimated volume of the microvessels, T is the distance between the sections sampled for the estimation, and A is the area of the cast profiles in m individual sections. When estimating the area *A*, the points hitting the profiles are counted, and their sum is multiplied by the area corresponding to each point. At least 200 points must be counted to obtain a reliable volume estimate (Gundersen & Jensen, 1987); see Fig. 14A.

To simulate the histological measurement of microvessel density Q_A (Fraser et al., 2012), i.e., the number of microvessel profiles *Q* per area unit *A*, an unbiased counting frame can be applied (Fig. 14B), and the microvessel density can be expressed using equation 5,

$$Q_A = \frac{Q}{A}. \tag{5}$$

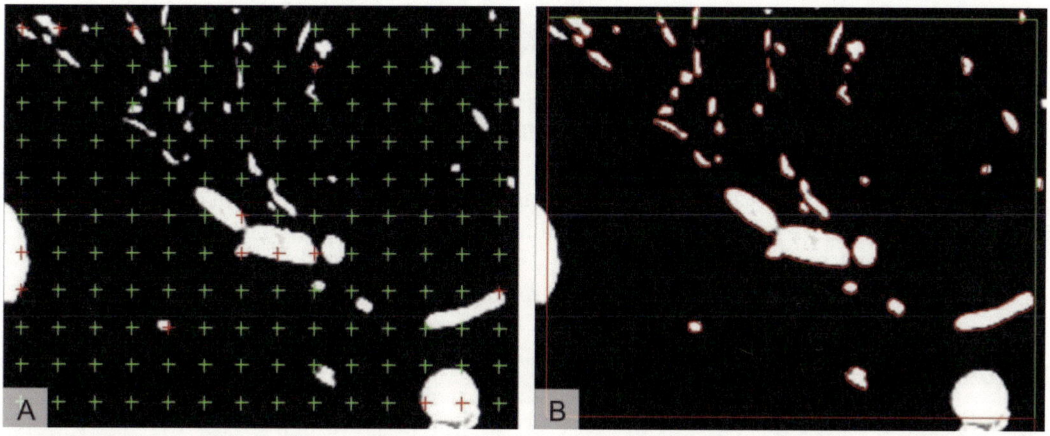

Figure 14. Estimating the volume fraction and the microvessel density. A - When estimating the area and volume of the microvascular cast, the points hitting the profiles (marked red) are counted, and their sum is multiplied by the area corresponding to each point. B – Counting the microvessel profiles per section area simulates the histological assessment of microvessel density. This procedure can be performed using the projection of an unbiased counting frame consisting of two admittance (green) and two forbidden (red) borders. Marked profiles of microvessels (red outlines) situated inside the frame or those touching admittance borders and not touching the forbidden lines are counted in the software Ellipse.

The surfaces of the microvascular casts can also be estimated using stereological methods. However, several of these methods require isotropic uniform random sections or vertical uniform random sections. Randomized orientation of the sections cannot always be guaranteed in micro-CT, as the sample is typically oriented with its long axis perpendicular to the X-ray beam (Fig. 1). The section plane is often arbitrary and cannot be regarded as random. A suitable solution without randomizing the cutting plane is using an isotropic virtual test probe named fakir (Larsen et al., 1998; Kubínová & Janáček, 1998; Kubínová & Janáček, 2001); see Fig. 15. The ratio between the surface area *S* and the reference volume *V(ref)* is called the surface density S_V, see equation 6:

$$S_V = \frac{S}{V(ref)}. \tag{6}$$

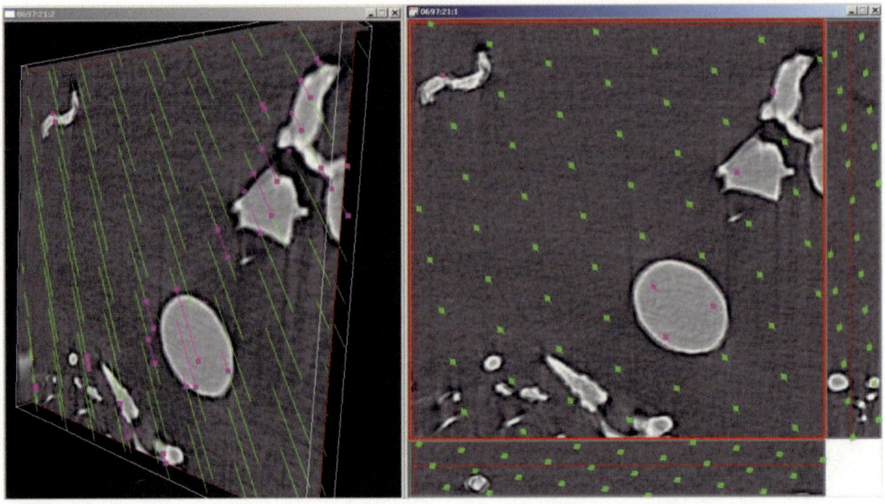

Figure 15. Estimating the surface in a series of micro-CT sections with arbitrary orientation using an isotropic triple spatial grid of orthogonal lines with a random initial orientation (fakir probe). The test lines of the fakir probe are green. The violet points denote intersections between the test lines and the current section. The left window shows a 3-D view, the right window shows the current section. Only one third of the triple line system is shown (software Ellipse).

8. Conclusion

This chapter reviews the current approaches in micro-CT imaging and the quantitative evaluation of the resulting image data sets. Both automatic image processing methods and interactive stereological methods are applied for the quantification of the volume fraction, surface density, length density, numerical density, orientation and anisotropy. Micro-CT imaging of bone tissue, tissue-engineered bone scaffolds, and microvascular corrosion casts is quantified using unbiased methods that are already acknowledged in quantitative histology.

Author details

Tomáš Gregor
New Technologies - Research centre, University of West Bohemia, Pilsen, Czech Republic

Petra Kochová
Department of Mechanics, Faculty of Applied Sciences, University of West Bohemia, Pilsen, Czech Republic

Lada Eberlová
Department of Anatomy, Faculty of Medicine in Pilsen, Charles University in Prague, Pilsen, Czech Republic

Lukáš Nedorost, Anna Králíčková and Milena Králíčková
Department of Histology and Embryology, Faculty of Medicine in Pilsen, Charles University in Prague, Pilsen, Czech Republic

Eva Prosecká
Laboratory of Tissue Engineering, Institute of Experimental Medicine, Academy of Sciences of the Czech Republic, v.v.i., Prague, Czech Republic
Institute of Biophysics, Charles University in Prague, Prague, Czech Republic

Václav Liška
Department of Surgery, Charles University Prague, University Hospital in Pilsen, Pilsen, Czech Republic

Hynek Mírka
Department of Imaging Methods, Faculty of Medicine in Pilsen, Charles University in Prague, Faculty Hospital in Pilsen, Pilsen, Czech Republic

David Kachlík
Department of Anatomy, Third Faculty of Medicine, Charles University in Prague, Prague, Czech Republic

Ivan Pirner and Petr Zimmermann
Department of Cybernetics, Faculty of Applied Sciences, University of West Bohemia, Pilsen, Czech Republic

Zbyněk Tonar
European Centre of Excellence NTIS - New Technologies for Information Society, Faculty of Applied Sciences, University of West Bohemia in Pilsen, Pilsen, Czech Republic

Acknowledgement

This work was supported by the European Regional Development Fund (ERDF) project "NTIS - New Technologies for Information Society", European Centre of Excellence, CZ.1.05/1.1.00/02.0090. The micro-CT technique was developed within the CENTEM project, reg. no. CZ.1.05/2.1.00/03.0088, which is cofunded from the ERDF within the OP RDI program of the Ministry of Education, Youth and Sports. The corrosion casting was funded by the Charles University in Prague, Project No. SVV 264808, and by the Internal Grant Agency of the Ministry of Health of the Czech Republic under Project No. IGA MZ ČR 13326. The quantification of vascular trees was funded by the Grant Agency of the Czech Republic, Project No. 106/09/0740. The bone scaffold research was funded by the Grant Agency of the Czech Republic, Project No. P304/10/1307, and by the The Grant Agency of the Charles University, Project GAUK No. 96610.

9. References

Badea, C. T.; Fubara, B.; Hedlund, L.W.; Johnson G. A. (2005). 4-D micro-CT of the mouse heart. *Mol. Imaging*, Vol. 4, pp. 110-116., ISSN 1535-3508.

Bag, S. ; Schambach, S.J. ; Boll, H. ; Schilling, L. ; Groden, C. ; Brochmann, M.A. (2010). Aktueller Stand der Mikro-CT in der experimentellen Kleintierbildgebung. *Fortschr. Roentgestr.*, Vol. 182, pp. 390-403, ISSN 1438-9029.

Bartel, H.; Lametschwandtner, A. (2000). Intussusceptive microvascular growth in the lung of larval Xenopus laevis Daudin: a light microscope, transmission electron microscope and SEM study of microvascular corrosion casts. *Anat. Embryol. (Berl).*, Vol. 202, pp. 55-65, ISSN 0340-2061.

Bentley, M.D.; Ortiz, M.C.; Ritman, E.L.; Romero, C. (2002). The use of Microcomputed tomography to study microvasculature in small rodents. *Am. J. Physiol.—Reg. Integ. Comp. Physiol.*, Vol. 282, pp. 1267–1279, ISSN 0363-6119.

Boll, H.; Nittka, S.; Doyon, F.; Neumaier, M.; Marx, A.; Kramer, M.; Groden, C.; Brockmann, M.A. (2011). Micro-CT based experimental liver imaging using a nanoparticulate contrast agent: a longitudinal study in mice. *PLoS One,* Vol. 6, e25692. ISSN 1932-6203.

Burghardt, A.J.; Link, T.M.; Majumar, S. (2011). High-resolution computed tomography for clinical paging of bone microstructure. *Clin. Orthop. Relat. Res.*, Vol. 469, pp. 2179-2193, ISSN 0009-921X.

Cancedda, R.; Cedola, A.; Giuliani, A.; Komlev, V.; Lagomarsino, S.; Mastrogiacomo, M.; Peyrin, F.; Rustichelli, F. (2007). Bulk and interface investigations of scaffolds and tissue-engineered bones by X-ray microtomography and X-ray microdiffraction. *Biomaterials,* Vol. 28, pp. 2505-2524, ISSN 0142-9612.

Dorsey, S.M.; Lin-Gibson, S.; Simon, C.G. Jr. (2009). X-ray microcomputed tomography for the measurement of cell adhesion and proliferation in polymer scaffolds. *Biomaterials,* Vol. 30, pp. 2967-2974, ISSN 0142-9612.

Fraser, G.M.; Milkovich, S.; Goldman, D.; Ellis, C.G. (2012). Mapping 3-D functional capillary geometry in rat skeletal muscle in vivo. Am. J. Physiol. Heart. Circ. Physiol., Vol. 302, pp. H654-664, ISSN 0363-6135.

Fujisaka, M.; Ohtani, O.; Watanabe, Y. (1996). Distribution of lymphatics in human palatine tonsils: a study by enzyme-histochemistry and scanning electron microscopy of lymphatic corrosion casts. *Arch. Histol. Cytol.*, Vol. 59, pp. 273-280, ISSN 0914-9465.

Gadžijev, E.M.; Ravnik, D. (1996). *Atlas of Applied Internal Liver Anatomy.* Springer, ISBN 978-3211827932, Wien.

Giuvărăşteanu, I. (2007). Scanning electron microscopy of vascular corrosion casts-standard method for studying microvessels. *Rom. J. Morphol. Embryol.*, Vol. 48, pp. 257-261, ISSN 0377-5038.

Gössl, M.; Rosol, M.; Malyar, N.M.; Fitzpatrick, L.A.; Beighley, P.E.; Zamir, M.; Ritman, E.L. (2003). Functional anatomy and hemodynamic characteristics of vasa vasorum in the walls of porcine coronary arteries. *Anat Rec A Discov Mol Cell Evol Biol.*, Vol. 272, pp. 526-537, ISSN 1552-4884.

Gundersen, H.J.; Jensen, E.B. (1987). The efficiency of systematic sampling in stereology and its prediction. *J. Microsc.*, Vol. 147, pp 229-263, ISSN 0022-2720.

Hajnis, K. (1988). The capacity of the sphenoid sinus. *Anat. Anz.*, Vol. 167, pp. 23-28, ISSN 0003-2786.

Havránková, J.; Skoda, V.; Holusa, R. (1989). The use of the Dentacryl rapid (Spofa) resin for preparation of the rat tracheobronchial casts. *Z. Versuchstierkd.*, Vol. 32, pp. 97-100, ISSN 0044-3697.

Ho, S.T.; Hutmacher, D.W. (2006). A comparison of micro CT with other techniques used in the characterization of scaffolds. *Biomaterials Vol.* 27, pp. 1362–1376, ISSN 0142-9612.

Hojo, T. (1993). Scanning electron microscopy of styrene-methylethylketone casts of the airway and the arterial system of the lung. *Scanning Microsc.*, Vol. 7, pp. 287-293, ISSN 0891-7035.

Howard, C.V.; Reed, M.G. (2005). *Unbiased Stereology: Three Dimensional Measurement in Microscopy*, (2nd edition), Garland Science/BIOS Scientific, ISBN 978-1859960899, New York.

Jorgensen, S.M.; Demirkaya, O.; Ritman, E.L. (1998). Three-dimensional imaging of vasculature and parenchyma in intact rodent organs with X-ray micro-CT. *Am. J. Physiol.*, Vol. 275, pp. H1103–1114, ISSN 0002-9513.

Kachlik, D.; Baca, V. (2006). Macroscopic and microscopic intermesenteric communications. *Biomed. Pap. Med. Fac. Univ. Palacky Olomouc Czech Repub.*, Vol. 150, pp. 121-124, ISSN 1213-8118.

Kachlík, D.; Hoch, J. (2008). *The blood supply of the large intestine.* Karolinum, pp. 68-71, ISBN 978-80-246-1397-0, Prague.

Kochová, P.; Cimrman, R.; Janáček, J.; Witter, K.; Tonar, Z. (2011). How to asses, visualize and compare the anisotropy of linear structures reconstructed from optical sections--a study based on histopathological quantification of human brain microvessels. *J. Theor. Biol.*, Vol. 286, pp. 67-78, ISSN 0022-5193.

Kubínová, L.; Janáček, J. (1998). Estimating surface area by the isotropic fakir method from thick slices cut in an arbitrary direction. *J Microsc.*, Vol. 191, pp. 201-211. ISSN 1365-2818.

Kubínová, L.; Janáček, J. (2001). Confocal microscopy and stereology: estimating volume, number, surface area and length by virtual test probes applied to three-dimensional images. *Microsc. Res. Techn.* Vol. 53, pp. 425-435. ISSN 1097-0029.

Lametschwandtner, A.; Minnich, B.; Kachlik, D.; Setina, M.; Stingl, J. (2004). Three-dimensional arrangement of the vasa vasorum in explanted segments of the aged human great saphenous vein: scanning electron microscopy and three-dimensional morphometry of vascular corrosion casts. *Anat. Rec. A Discov. Mol. Cell. Evol. Biol.*, Vol. 281, pp. 1372-1382, ISSN 1552-4884.

Lametschwandtner, A.; Minnich, B.; Stöttinger, B.; Krautgartner, W.D. (2005). Analysis of microvascular trees by means of scanning electron microscopy of vascular casts and 3D-morphometry. *Ital. J. Anat. Embryol.*, Vol. 110, pp. 87-95, ISSN 1122-6714.

Larsen, J.O.; Gundersen, H.J.; Nielsen, J. (1998). Global spatial sampling with isotropic virtual planes: estimators of length density and total length in thick, arbitrarily orientated sections. *J. Microsc.*, Vol. 191, pp. 238-248 ISSN 1365-2818.

Lee, T.C.; Kashyap, R. L.; Chu C.N. (1994). Building skeleton models via 3-D medial surface/axis thinning algorithms.*Graph. Model. Image Process.*, Vol. 56, No. 6, pp. 462-478, ISSN 1077-3169.

Lokkegaard, A.; Nyengaard, J.R.; West, M.J. (2001). Stereological estimates of number and length of capillaries in subdivisions of the human hippocampal region. *Hippocampus*, Vol. 11, pp. 726-740, ISSN 1050-9631.

Ma, X.; Tian, J.; Yang, X.; Qin, C. (2011): Molecular paging in tumor angiogenesis and relevant drug research. *Int. J. of Biomedical Imaging*, Vol. 2011, 370701. ISSN 1687-4196.

Marques-Sampaio, B.P.; Pereira-Sampaio, M.A.; Henry, R.W.; Favorito, L.A.; Sampaio, F.J. (2007). Dog kidney: anatomical relationships between intrarenal arteries and kidney collecting system. *Anat Rec. (Hoboken)*, Vol. 290, pp. 1017-1022, ISSN 1932-8486.

Marsen, M.; Paget, C.; Yu, X.; Henkelman, R.M. (2006). Estimating perfusion using micro-CT to locate microspheres. *Phys. Med. Biol.*, Vol. 51, pp. N9-16, ISSN 0031-9155.

Mather, M.L.; Morgan, S.P.; Crowe, J.A. (2007). Meeting the needs of monitoring in tissue engineering. *Regen. Med.*, Vol. 2, pp. 145–160, ISSN 1746-0751.

Mather, M.L.; Morgan, S.P.; White, L.J.; Tai, H.; Kockenberger, W.; Howdle, S.M. (2008). Image-based characterization of foamed polymeric tissue scaffolds. *Biomed. Mater.*, Vol. 3, pp. 015011 (11pp), ISSN 0955-7717.

Minnich, B.; Bartel, H.; Lametschwandtner, A. (2002). How a highly complex three-dimensional network of blood vessels regresses: the gill blood vascular system of tadpoles of Xenopus during metamorphosis. A SEM study on microvascular corrosion casts. *Microvasc Res.*, Vol. 64, pp. 425-437, ISSN 0026-2862.

Missbach-Guentner, J.; Hunia, J.; Alves, F. (2011). Tumor blood vessel visualization. *Int. J. Dev. Biol.*, Vol. 55, pp. 535-546, ISSN 1696-3547.

Morris, D.E.; Mather, M.L.; Crowe, J.A. (2009). Generation and simulated imaging of pseudo-scaffolds to aid characterisation by X-ray micro CT. *Biomaterials*, Vol. 30, pp. 4233–4246, ISSN 0142-9612.

Mouton, P.R. (2002). *Principles and Practices of Unbiased Stereology. An Introduction for Bioscientists,* The Johns Hopkins University Press, ISBN 0-8018-6797-5, Baltimore.

Müller, R. (1994). Non-invasive bone biopsy: a new method to analyse and display the 3D structure of trabecular bone. *Phys. Med. Biol.*, Vol. 39, pp. 145–164, ISSN 0031-9155.

Müller, R.; Hildebrand, T.; Hauselmann, H.J.; Rüegsegger, P. (1996). Resolution dependency of microstructural properties of cancellous bone based on 3d-microct. *Technol. Health Care,* Vol. 4, pp. 113–119, ISSN 0928-7329.

Müller, R.; van Campenhout, H.; van Damme, B.; van der Perre, G.; Dequeker, J.; Hildebrand, T.; Rüegsegger, P. (1998). Morphometric analysis of human bone biopsies: a quantitative structural comparison of histological sections and micro-computed tomography. *Bone*, Vol. 23, pp. 59–66, ISSN 8756-3282.

Müller, R. (2009). Hierarchical microimaging of bone structure and function. *Nat. Rev. Rheumatol.*, Vol. 5, pp. 373-381, ISSN 1759-4790.

Paluzzi, A.; Belli, A.; Bain, P.; Viva, L. (2007). Brain 'imaging' in the Renaissance. *J. R. Soc. Med.*, Vol. 100, pp. 540-543, ISSN 0141-0768.

Peyrin, F. (2011). Evaluation of bone scaffolds by micro-CT. *Osteoporos. Int.*, Vol. 20, pp. 2043-2048, ISSN 0937-941X.

Prosecká, E.; Rampichová, M.; Vojtová, L.; Tvrdík, D.; Melčáková, Š.; Juhasová, J.; Plencner, M.; Jakubová, R.; Jančář, J.; Nečas, A.; Kochová, P.; Klepáček, K.; Tonar, Z.; Amler, Z. (2011). Optimized conditions for mesenchymal stem cells to differentiate into osteoblasts on a collagen/hydroxyapatite matrix. *J. Biomed. Mater. Res. A,* Vol. 99A, pp. 307-315, ISSN 1552-4965.

Razavi, H.; Dusch M.N.; Zarafshar, S.Y.; Taylor, C.A.; Feinstein, J.A. (2012). A method for quantitative characterization of growth in 3-D structure of rat pulmonary arteries. *Microvasc. Res.,* Vol. 83, pp. 146-153, ISSN 0026-2862.

Renghini, C.; Komlev, V.; Fiori, F.; Verné, E.; Baino, F.; Vitale-Brovarone, C. (2009). Micro-CT studies on 3-D bioactive glass-ceramic scaffolds for bone regeneration. *Acta Biomater.,* Vol. 5, pp. 1328-1337, ISSN 1742-7061.

Ritman, E.L. (2011). Current status of developments and application of micro-CT. *Annu. Rev. Biomed. Eng.,* Vol. 13, pp. 531-552, ISSN 1523-9829.

Roth, H.; Neubrand, T.; Mayer, T. (2010). Improved inspection of miniaturised interconnections by digital X-ray inspection and computed tomography, *Proceedings of Electronics Packaging Technology Conference (EPTC),* ISBN 978-1-4244-8560-4, Singapore, December 2010.

Rüegsegger, P. (2001). Imaging of bone structure, In: *Bone Mechanics Handbook,* 2nd ed., Cowin S.C. (Ed.), pp.9/1-9/24, CRC Press, ISBN 978-0849391170, Boca Raton.

Savai, R.; Langheinrich, A.C.; Schermuly, R.T.; Pullamsetti, S.S.; Dumitrascu, R.; Traupe, H.; Rau, W.S.; Seeger, W.; Grimminger, F.; Banat, G.A. (2009). Evaluation of angiogenesis using micro-computed tomography in a xenograft mouse model of lung cancer. *Neoplasia,* Vol. 11, pp. 48-56, ISSN 1522-8002.

Schambach, S.J.; Bag, S.; Schilling, L.; Groden, Ch.; Brockmann, M.A. (2010). Application of micro-CT in small animal paging. *Methods,* Vol. 50, pp. 2-13 ISSN 1046-2023.

Schladitz, K. (2011). Quantitative micro-Ct. *J. Microscopy,* Vol. 243, pp. 111-117, ISSN 1365-2818.

Sonka, M.; Hlavac, V.; Boyle, R. (1998). Image processing, Analysis, and Machine Vision. *PWS Publishing,* pp. 123-134, 232-235, ISBN 0-534-95393-X.

Sterio, D.C. (1984). The unbiased estimation of number and sizes of arbitrary particles using the disector. *J. Microsc.,* Vol. 134, pp. 127–136, ISSN 1365-2818.

Tjia, J.S., Moghe, P.V. (1998). Analysis of 3-D microstructure of porous poly(lactide-glycolide) matrices using confocal microscopy. *J. Biomed. Mater. Res.,* Vol. 43, pp. 291–299, ISSN 1552-4965.

Tonar, Z.; Khadang, I.; Fiala, P.; Nedorost, L.; Kochová, P. (2011). Quantification of compact bone microporosities in the basal and alveolar portions of the human mandible using osteocyte lacunar density and area fraction of vascular canals. *Ann Anat.,* Vol. 193, pp. 211-219, ISSN 0940-9602.

Villa, C.; Erratico, S.; Razini, P.; Fiori, F.; Rustichelli, F.; Torrente, Y.; Belicchi, M. (2010). Stem cell tracking by nanotechnologies. *Int J. Mol. Sci.,* Vol. 11, pp. 1070-1081, ISSN 1422-0067.

Zagorchev, L.; Oses, P.; Zhuang, Z.; Moodie, K.; Mulligan-Kehoe, M.J.; Simons, M.; Couffinhal, T. (2010). Micro computed tomography for vascular exploration. *J. Angiogen. Res.*, Vol. 2, pp 7-18, e-ISSN 2040-2384

Chapter 11

Mechanical Behavior of Articular Cartilage

Nancy S. Landínez-Parra, Diego A. Garzón-Alvarado
and Juan Carlos Vanegas-Acosta

Additional information is available at the end of the chapter

http://dx.doi.org/10.5772/46881

1. Introduction

Articular Cartilage (AC) is a poro-elastic biological material that allows the distribution of mechanical loads and joint movements. As a biphasic material, in the presence of load, the articular cartilage deforms its solid matrix and modifies the fluid hydrostatic pressure within. The aim of this chapter is to present a mathematical model that predicts the mechanical behavior of articular cartilage, taking into account the duality between the solid matrix and articular liquid, and its poro-elastic characteristics. Using a finite element method approach, the response of a piece of articular cartilage in one and two dimensions has been simulated, with tensile, compressive and oscillatory mechanical loads. The analysis of the results allows a qualitative validation of the poro-elastic behavior of the model due to the solid matrix deformation and the fluid outflow that causes variations of pressures inside the articular cartilage in accordance with reported trials. The mathematical model allows for prediction of articular cartilage's biomechanical behavior. These results contribute to the research processes in fields of study such as biomechanics and tissue engineering.

2. Background

One of the pathologic entities that most often affect quality of life of individuals is osteoarthrosis (OA), which is caused by the deterioration of cartilage in synovial joints. In the U.S. in the early nineties it was estimated that 37.9 million people (which constituted 15% of the population) suffered from one of the various existing musculoskeletal diseases. OA was present in 21 million individuals (Lawrence et al., 1998).

OA compromises skeletal muscle function, causing pain and difficulty in basic activities of daily living. Several studies have shown that the forces exerted on cartilage can modify its structure and composition, resulting in a change in the biomechanical behavior of the same (Wu & Kirk, 2001). The onset and progression of OA are commonly affected by mechanical factors associated with either joint loading or local contact stress (Andriacchi et al., 2004).

The mechanical environment of cartilage cells (chondrocytes) is an important factor influencing joint health and function. Chondrocytes in articular cartilage utilize mechanical signals in conjunction with other environmental, genetic, extrinsic and intrinsic hormonal and/or paracrine or autocrine factors to regulate their metabolic activity. This capability provides the means by which articular cartilage may alter its structure and composition to meet the physical demands of the body (Boschetti et al., 2004).

The permeability of cartilage contributes to many tissue functions such as transport of food to the chondrocytes, the ability to withstand high loads, and maintaining a fluid film to lubricate opposing joint surfaces (Guilak et al., 1999). Measuring the cartilage permeability considering its depth, by the behavior of a fluid flow induced by the application of a pressure gradient, can show a decrease in permeability in relation to depth for each level of the applied pressure difference.

Here is a review of the anatomy, morphology and physiology of articular cartilage for the purpose of more clearly understanding its responses to load and its relationship to the deformation processes and the changes of fluid pressure within.

2.1. Biological tissues

Biological materials are generally multiphase, coexisting in a solid and a fluid phase (Hubertus Frijns, 2000; Doblaré, 2005; Haider & Schugart, 2006). They also have a strong microstructure that gives them a clear heterogeneous and anisotropic condition and, in addition, its mechanical behavior is strongly nonlinear nonlinear (Wilson et al., 2004; Ateshian et al., 1997; Chan et al., 2004). All of this, without taking into account important aspects such as the strong dependence on age, sex, metabolism, and in particular history and diseases and, ultimately, cell activity and its interaction with the particular environment in which it develops.

In this sense, it can be seen as soft biological tissues in which the elastic modulus is approximately equal to the stresses to which they are subjected. Some typical cases correspond to the arteries and veins, cartilage, ligaments, tendons, muscles or skin. In general, they are composites, made up of organic matrix reinforced by fibers of collagen and elastin. Its behavior again depends on its structural composition, especially the percentage of fibers, their characteristics and the type of grouping. Thus, tissues specialized for tensile strength (e.g. ligaments) are rich in fiber and their orientation essentially coincides with the direction of stress to which they are subjected, while the elastic absorbing the compressive forces (e.g. cartilage) are rich in proteoglycans and fibers distributed in various directions. The cartilage then can be referred to as hydrated tissue, which has a highly compressible behavior corresponding to a saturated biphasic material (if one considers the four-phase ion diffusion) with the possibility of the escape of the fluid inside. The main purpose of this behavior is to provide optimum synovial joint lubrication conditions, friction, wear, shock absorption and load distribution. Interstitial fluid flow in these tissues is crucial both in their viscoelastic properties as well as in the lubrication mechanism; contemplating this is necessary to model the behavior of such tissues. Furthermore, they are again heterogeneous and anisotropic due to the preferred orientation of collagen fibers (Wilson et al., 2004).

Understanding the importance of in-depth knowledge of the composition and behavior of articular cartilage, in the next section this tissue will be described.

2.2. Articular cartilage

Cartilage is categorized as an avascular, aneural and alinfatic tissue. It's composed of cells called chondrocytes, surrounded by an extracellular matrix (ECM) that they secrete. It's formed by an abundant extracellular matrix in which the chondrocytes are located in spaces called gaps (See Fig 1). The chondrocytes synthesize and secrete the organic components of the extracellular matrix, which are essentially collagen, hyaluronic acid, proteoglycans and glycoproteins. Hyaline and fibrous cartilage are distinguished by the characteristics of the matrix. There is also elastic cartilage, in which the elastin is a part of the extracellular matrix (Koenig, 2011; Sopena-Juncosa et al., 2000; Nordin & Frankel, 2004).

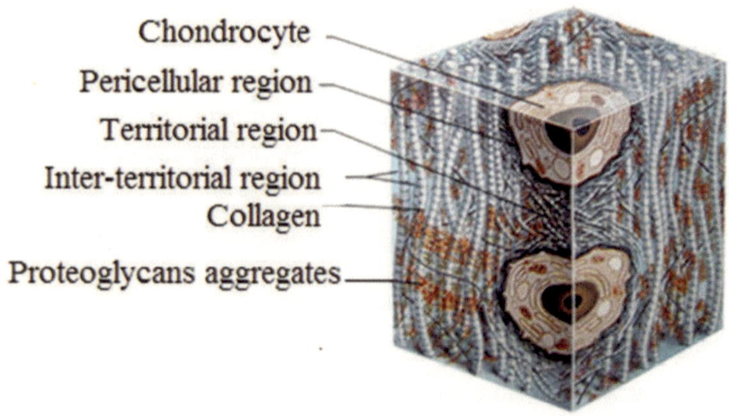

Figure 1. Representation of a fragment of Articular Cartilage. (Sopena-Juncosa et al., 2000)

For a better understanding, each of the structures mentioned above will be defined. Initially chondrocytes will be described, then subsequently the ECM and each of its components.

2.2.1. Chondrocytes

They are cells sparsely distributed in the tissue. They constitute approximately 10 percent of cartilage volume. They are located in some gaps within the ECM, to which they adapt. They produce the adjacent ECM, but they are also capable of depolymerizing and removing it to broaden its gaps. (This fact is very evident in the process of endochondral ossification) The key feature of the intercellular substance is its hyper-hydration state (80 percent), and combined with proteoglycans, it forms a gel. Greater or lesser elasticity of the cartilage depends on its water content (Koenig, 2011; Sopena-Juncosa et al., 2000; Nordin & Frankel, 2004).

2.2.2. Extracellular Matrix - ECM

This consists of an organized and dense network of thin collagen fibers embedded in a concentrated solution of proteoglycans. It is responsible for the mechanical properties of cartilage (Koenig, 2011; Sopena-Juncosa et al., 2000; Nordin & Frankel, 2004). This matrix is composed of:

Water: (60-80 percent) Water is the main component of cartilage, which contributes to its damping properties, cartilage nutrition and the articular lubrication processes. It allows for the deformation of the cartilage in response to mechanical loads, flowing inside and outside them.

Collagen: (10-20percent). Predominantly type II (90-95 percent), giving the cartilage great tensile strength.

Proteoglycans: (PGs) (10-15 percent). These are complex macromolecules, responsible for the resistance to compression of the cartilage. They are secreted by chondrocytes and composed of subunits called glycosaminoglycans (GAGs). The most common GAG is chondroitin-sulfate (of which there are 2 subtypes, the chondroitin-4-sulfate and the chondroitin-6-sulfate), then the keratan-sulphate (or keratin-sulfate) and the dermatan-sulphate. Chondroitin-4-sulphate is the most abundant and decreases over the years; chondroitin-6-sulfate remains constant; and keratan-sulphate increases with age. The PGs have an average lifespan of three months and have a great capacity for retain water which gives elasticity to the tissue. They are attached to collagen and are responsible for the "porous" structure of cartilage (See Fig.1)

Extracellular Glycoproteins: (anchorite CII, fibronectin, laminin, integrin). They serve a binding function between the ECM and chondrocytes. The most important, integrin, interacts with cell receptors and regulates the migration, proliferation and differentiation of the chondrocytes (Sopena-Juncosa et al., 2000).

At present it is believed that collagen has a different orientation in articular cartilage in relation to its depth, as described by Benninghoff in 1925 in the Arcade model (Wilson, 2005): Packets of primary fibrils extend perpendicular to the sub-chondral bone; the fibrils are separated near the joint surface presented by the curve of the horizontal; each packet of the vertical surface is assumed to be subdivided into two different directions in the curvature of the radial direction (See Fig. 2). It was assumed that the orientation of the secondary fibrils is random and that on the uppermost, fibers are distributed horizontally.

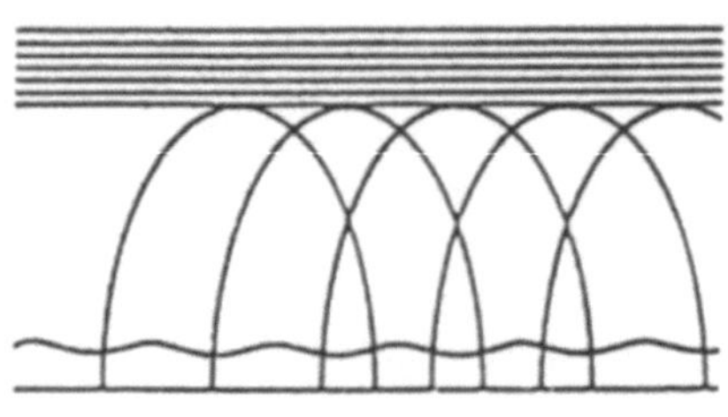

Figure 2. Representation of collagen distribution in articular cartilage.

2.3. Architecture of articular cartilage

Some studies like those reported by Martin, 2002; Sopena et al., 2000; Wilson, 2005 and Meyer & Wiesmann, 2010; mention that with electron microscopy it can be determined that cartilage has a multilayered arrangement with layers of different thickness and composed of fibrils irregularly cross-linked into a plane parallel to the surface. Using the electron microscope, the following layers of articular cartilage were described:

2.3.1. Superficial, tangential or sliding layer

This is adjacent to the joint cavity. The chondrocytes adopt an elongated or ellipsoid shape and they are oriented parallel to the surface. The cells have low activity and poor protein synthesis. This layer possesses few PGs and a high concentration of fine collagen fibers distributed perpendicularly to one another and parallel to the surface in order to withstand the shear forces during joint movement.

2.3.2. Intermediate or transitional layer

Cells adopt a rounded morphology and are larger than those of the previous layer. The chondrocytes are irregularly arranged and show a greater presence of PGs and less collagen with thicker fibers arranged obliquely and randomly in all three planes of space. This layer has high metabolic activity and supports compression forces.

2.3.3. Radial or deep layer

The cells are rounded and have the same characteristics as layer 2 but adopt a columnar arrangement. They present a high protein synthesis. The collagen fibers are thick and they are distributed parallel to each other and perpendicular to the articular surface to provide resistance to compressive forces. The water content is less than in the previous layers and proteoglycans are most abundant.

2.3.4. Calcified layer

This is adjacent to the bone and separated from the previous layer by a basophilic line called tidal or "tidemark", which is a bar wavy tangential arrangement of its fibers and can withstand shear forces. The cells are small and scarce. The matrix is rich in hydroxyapatite crystals. Cartilage anchoring to the sub-chondral bone occurs in this layer (See Fig 3).

The most common tests used for explaining the behavior of articular cartilage under load, expressed in computer models, that include the behavior of swelling or anisotropic properties of the collagen structure for determining the mechanical quality of articular cartilage are: confined compression, the unconfined compression, indentation and swelling (Wilson, 2005). This can be reviewed extensively in the literature.

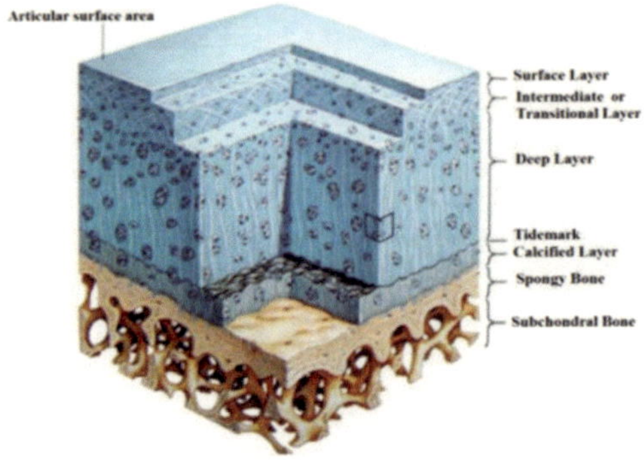

Figure 3. Architectural layout of the articular cartilage according to its various layers. Note the anisotropic distribution of the tissue in relation to the depth thereof. (Sopena-Juncosa et al., 2000)

2.4. Biphasic behavior of articular cartilage

Mechanical properties of articular cartilage are attributed to their complex structure and composition of the extracellular matrix that is comprised of a fluid phase (water containing dissolved ions) and a solid matrix that consists mainly of a fibrous network of collagen type II and aggregates of proteoglycans as well as other type of proteins, lipids, and cells (Wilson et al., 2004).

With the mechanical load, the interstitial fluid is redistributed through the pores of the permeable solid matrix, resulting in predominantly viscoelastic behavior (See Fig 4). This highly viscoelastic behavior of articular cartilage is mainly due to two mechanisms: (a) the frictional drag force of interstitial fluid flow through the porous solid matrix (i.e., the flow-dependent mechanism), and (b) the function of the time-dependent deformability of the solid matrix (i.e. the flux-independent mechanism) (Garzón, 2007).

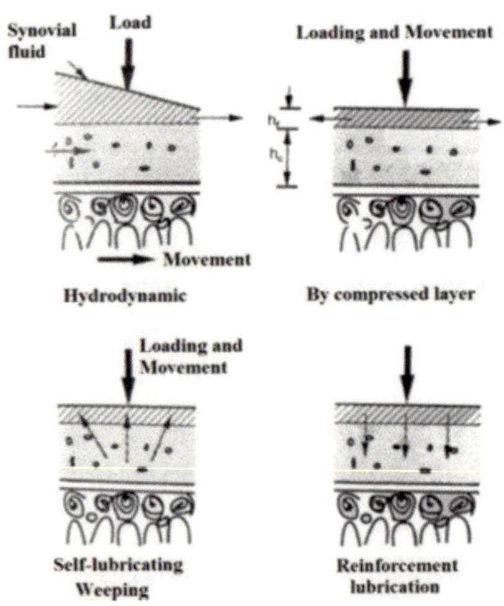

Figure 4. Different forms of lubrication in articular cartilage given by the load applied to the tissue. This lubrication and nutrition takes various forms but mainly by the displacement of the fluid. (Terada et al., 1998).

Mechanical behavior of cartilage is then described by the biphasic or continuous poro-elastic model that describes the mechanical interactions of the different phases (Donzelli & Spilker, 1998).

From the mechanical standpoint, the most important components of articular cartilage are strong and highly organized as a network of collagen together with the load of proteoglycans. Due to the fixed charges of proteoglycans, the cation concentration within the tissue is higher than in the surrounding synovial fluid. This excess of ion particles leads to an osmotic pressure difference, which causes swelling tissue. The fibrillar collagen network resists pressure and swelling. This combination makes cartilage a unique, highly hydrated and pressurized tissue, reinforced by the tension of the collagen network (Wilson, 2005).

2.4.1. Mixtures theory

Articular cartilage can be described by the mixtures theory as a mixture of four elements: a fibrous network (collagen fibres and proteoglycans), a fluid and a positively and negatively charged particle. However, it's important to differentiate between components and phases. Hubertus, 2000; defines a component as a group of particles with the same properties and a phase as a set of miscible components. Thus in theory the four components can be separated into only two phases: a solid and a fluid phase. In this case the fluid is comprised of three components: the liquid, the cations and anions.

Many authors such as Haider & Schugart, 2000; Wilson et al., 2004; Haider & Guilak, 2007; Meng et al., 2002; Wu et al., 1997; Terada et al., 1998; Donzellie et al., 1999 and Donzelli & Spilker, 1998; among others, have conducted their research, viewing cartilage from this biphasic behavior. This has allowed its analysis as a material with a porous-viscoelastic behavior, in an attempt to better understand its response to loads, forces and overload.

2.5. Computational analysis of articular cartilage

Continued use of simulation in medicine has allowed important data to be obtained about the biological, mechanical and chemical behavior of the organs and tissues using mathematical formalization and subsequent numerical simulation of complex biological processes. Various medical problems related to surgery, trauma and rehabilitation have been identified, conceptualized and solved systematically and numerically (Garzón, 2007).

Computer mechano-biology determines the quantitative rules governing actions for cellular expression, differentiation and maintenance of biological and mechanical stimuli, which can be simulated by numerical methods. The procedure for finding such rules is usually through the process of "trial and error" (Van der Meulen & Huiskes, 2002). The computational tests are simulated usually from problems in the contour value by which the mechanical loads on the boundary are transferred to local mechanical variables (stress and strain). On the biological side, these local mechanical or biophysical variables stimulate cell expression to regulate, for example, the composition of the matrix and the expression of molecular substances. Both biological and mechanical parts are combined in a computational model, which considers the application of forces, mechano-transduction, cellular expression, genetics and the transformation of the characteristics of the extracellular matrix. The typical method of numerical implementation of these mechano-biological problems is the finite element method (FEM) (Garzón, 2007).

Finite element computational analysis has been used as an approach to diverse biological processes including the biomechanical behavior of articular cartilage (Wilson et al., 2004; Ateshian et al., 1997; Chan et al., 2004; Wilson, 2005; Donzelli et al, 1999; Donzelli & Spilker, 1998; Almeida & Spilker, 1998; Wu & Herzog, 2000; Levenston et al., 1998). Using the material representation of the continuum phase of cartilage, results have indicated that local intermittent hydrostatic pressure promotes cartilage maintenance (Carter & Wong, 2003).

3. Mechanical behavior of articular cartilage (AC):

Mechanical properties of articular cartilage (AC) are attributed to its complex structure and to composition of its ECM including a fluid phase (water with dissolved ions), and a solid matrix (collagen type II, aggregates of PGs, proteins, lipids, and cells) (Haider & Guilak, 2007). With the mechanical load, the interstitial fluid is redistributed through the pores of the permeable solid matrix, resulting in predominantly poro-elastic conduct. This behavior of the AC is mainly due to two mechanisms: (a) the frictional force due to drag flow of interstitial fluid through the porous solid matrix (flow-dependent mechanism), and (b) the deformability of the matrix strong function of time (flow-independent mechanism) (Wilson, 2005).

4. Mathematical model for articular cartilage

Several authors (Mow, 1980 ; Haider & Schugart, 2006; Wilson et al., 2005; Haider & Guilak, 2007; Meng et al., 2002; Wu et al., 1997; Terada et al., 1998; Donzelli et al., 1999; Donzelli & Spilker, 1998); have conducted their research on cartilage from biphasic behavior that this tissue exhibits. This allowed analysis of the same material as a poro-elastic behavior capable of supporting loads. The mathematical model of the AC as a biphasic material, analyzes the displacement *u (t, x)* of the solid (matrix) and pressure *p (x)* of the fluid displaced by the load, thanks to its characteristic of poro-elasticity. This model is described by the equations (1) and (2):

$$-\nabla.(2\mu_s \underline{\underline{\varepsilon}}(\underline{u}) + \lambda_s \nabla.\underline{u}\,\underline{\underline{I}}) + \nabla p = 0 \text{ en } \Omega \tag{1}$$

$$\frac{\partial}{\partial t}(\nabla.\underline{u}) - \nabla.(k\nabla p) = 0 \text{ en } \Omega \tag{2}$$

Equation (1) is derived from the law of conservation of momentum and corresponds to the linear elasticity equation (term 1a) coupled with a term that represents fluid pressure (term 1b). The term $\underline{\underline{\varepsilon}}(\underline{u})$ corresponds to the strain tensor acting on the Γsurface enclosed byΩ. μ_s y λ_s are the "Lame elastic constants" for the solid, related to Young's modulus and

Poisson's ratio (E, ν). For its part, the equation (2) refers to the change of the dilation of the solid matrix (term 2a) due to the mechanical load created by the divergence of the gradient of the pressure of fluid contained in the domain Ω (term 2b) (Frijns, 2000). In this equation, k is a constant representing the permeability of solid module.

4.1. Boundary conditions

Boundary conditions of the model are defined in Γ domain and may be dependent on time. The mathematical expression of these conditions is (3-6):

$$u = g_u^D \text{ en } \Gamma_u^D \tag{3}$$

$$n.(2\mu_s \underline{\underline{\varepsilon}}(\underline{u}) + \lambda_s \nabla.\underline{u}\underline{\underline{I}} - p\underline{\underline{I}}) = g_u^N \text{ en } \Gamma_u^N \tag{4}$$

$$p = g_p^D \text{ en } \Gamma_p^D \tag{5}$$

$$-n.(k\nabla p) = g_p^N \text{ en } \Gamma_p^N \tag{6}$$

$$\text{with } \Gamma = \Gamma_u^P + \Gamma_u^N + \Gamma_p^D + \Gamma_p^N$$

A widely used method for solving partial differential equations in complex geometries is the finite element method (Garzón, 2007). This method allows implementing the numerical model presented in equations (1) and (2) simply and with low computation cost. The method consists of using a vectorial function *W* or *weighting function* and a scalar function of *q*, which minimizes the terms of the equations (1) and (2). Multiplying (1) by *W* and (2) by *q*, and performing integration by parts in the Ω domain, we obtain a variational of the form (Frijns, 2000):

$$a(u,W) + b(W,p) = (g_u^N, W) \tag{7}$$

$$\frac{\partial}{\partial t} b(u,q) - c(p,q) = (g_p^N, q) \tag{8}$$

where:

$$a(u,W) = \int_\Omega (2\mu_s \underline{\underline{\varepsilon}}(\underline{u}) : \varepsilon(W) + \lambda_s (\nabla.u)(\underline{\nabla}.W) \tag{9}$$

$$b(W,q) = -\int_\Omega \nabla.Wq \tag{10}$$

$$c(p,q) = \int_\Omega (k\nabla p).\nabla q \tag{11}$$

$$(g_u^N, W) = \int_{\Gamma_u^N} g_u^N.W \tag{12}$$

The solution method is based on dividing the domain Ω (continuum) in which are defined equations (1) and (2) in their integral form in a series of non-intersecting subdomains Ω^e called *finite elements*. The set of finite elements forms a partition of the domain called *discretization* which together are described by equations (14) and (15):

$$a(u_u^h, W_h^h) + b(W_h^h, p_n^h) = (g_u^N, W_h^h) \tag{14}$$

$$b(u_n^h, q^h) - \Delta t\, c(p^h, q_n^h) = b(u_{n-1}, q^h) + \Delta t(g_p^N, q^h) \tag{15}$$

In (15) the $n-1$ subscript indicates the value of a parameter at a $t-1$ time, while the subscript n indicates the value in the next time, $t_n = t_{n-1} + \Delta t$ (Frijns, 2000). Using the Galerkin method, the functions *W* and *q* are approximated by the matrix expression (16) where the ***N*** vector contains *shape functions standard* (ξ, η) which allow interpolation within each element of the domain.

$$q^h = W_h^h = \begin{pmatrix} N_1 \\ N_2 \\ N_3 \end{pmatrix} = \begin{pmatrix} \frac{1}{2}\xi(\xi - 1) \\ (1+\xi)(1-\xi) \\ \frac{1}{2}\xi(\xi + 1) \end{pmatrix} \tag{16}$$

$$W_h^{\ h} = \begin{pmatrix} N_1 \\ N_2 \\ N_3 \\ N_4 \end{pmatrix} = \begin{pmatrix} \frac{1}{4}(1-\xi)(1-\eta) \\ \frac{1}{4}(1+\xi)(1-\eta) \\ \frac{1}{4}(1+\xi)(1+\eta) \\ \frac{1}{4}(1-\xi)(1+\eta) \end{pmatrix} \tag{17}$$

In (16), the shape functions correspond to the case of one-dimensional element with three nodes (See Fig. 5a). For two-dimensional case four node elements are used, whose standard shape functions are shown in matrix form (17) and in Figure 5b.

If the integration space *(x,y)* is changed to $(\xi,\ \eta)$ by the *Jacobian* of the *transformation* and vector notation is adjusted, the equations (1) and (2) can be reduced to a matrix system of elementary type of the form (18) which corresponds to the algebraic discretization in Ωe domain of an element, with *k* elementary stiffness matrix, *m* the unknowns and *f* the independent term. Joining the result of (18) for total elements in Ω, a general matrix system is obtained defined as (19) where *K* is the global matrix of stiffness, *M* is the vector of unknowns and *F* is the global vector entries.

$$k \cdot m = f \text{ en } \Omega^e \tag{18}$$

$$K \cdot M = F \text{ en } \Omega \tag{19}$$

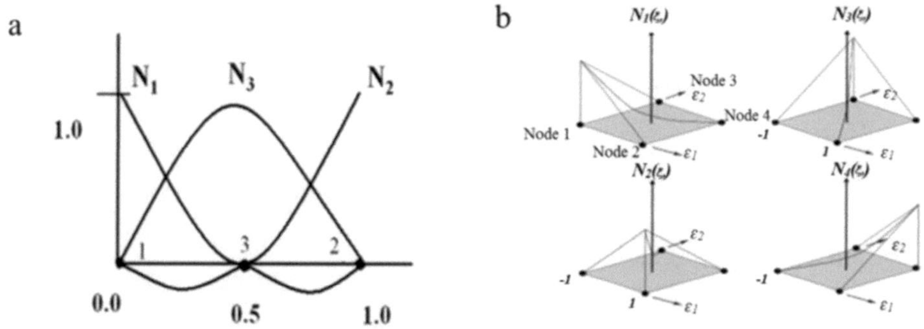

Figure 5. Representation of the *shape functions standard.* a) Case 1D, 3 nodes per element. b) Case 2D elements with four nodes (Radi, 2007).

4.2. Model implementation

Solution of these equations for both u and p are implemented using a routine programmed in Fortran and a desktop PC with 2.4 GHz AMD processor and 1.0 GB of RAM

4.3. Computer simulation

Conditions for the simulation were based on experiments by Frijns, 2000 and Wu et al., 1999. Simulations were performed for 1D and 2D looking tissue response to the application of: (a) compression, (b) tensile and (c) oscillating or cyclical loads. The simulation time is 45 seconds of load in all cases. The loads applied in 1D and 2D simulations were performed with the parameters shown in Table 1. The tests considered the AC as a continuous and homogeneous material with the chondrocytes being part of the continuum.

	E	ν	K
Matrix	0.40 Mpa	0.1	1.0 x10^{-2}mmN^{-1}s^{-1}
Chondrocyte	1.0 kPa	0.2	1.0 x10^{-5}mmN^{-1}s^{-1}

Table 1. Parameters for the homogeneous material characteristics (Wu et al., 1999).

4.3.1. Meshing

For the 1D simulation a mesh was performed that represents a fragment of 0.3 mm AC. From this mesh 641 nodes and 320 elements of 3 nodes were obtained. In the 2D simulation a mesh was performed that represents a fragment of articular cartilage 0.09 mm2 (0.3 mm x 0.3 mm). In this case 10,201 nodes were obtained, equivalent to 10,000 elements of 4 nodes.

4.3.2. Boundary conditions

Simulation was performed so as not to allow displacement at the bottom. Load was applied on the upper edge, allowing the fluid outlet only at the bottom of the tissue fragment, as shown in Figure 6a. For 2D, a condition of tissue confinement was simulated, as shown the Figure 6b, so as to present only flow at the bottom. The burden was placed at the top and lateral and bottom movement were restricted, similar conditions to those reported in several experimental studies, including that of Ateshian et al, 1997; Frijns, 2000 and Wu et al, 1999.

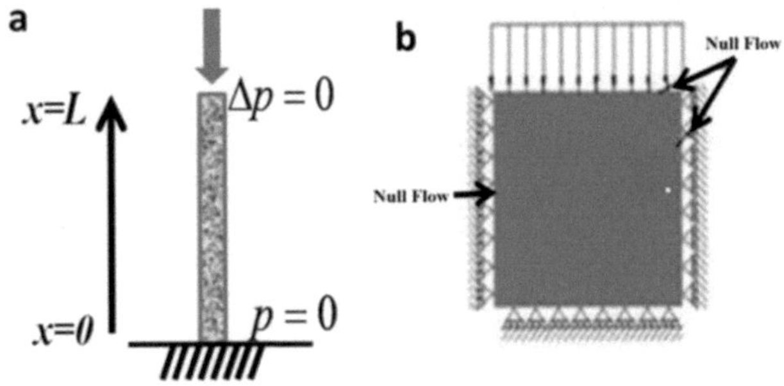

Figure 6. AC scheme for the confinement conditions in (a) 1D and (b) 2D.

4.3.3. Loading conditions

Calculations to simulate the applied load were performed from the data for the AC reported by Wu et al. For the 1D simulations, a load was applied of -2.033 N/m in compression, 2.033 N/m in tension and cyclic loading with frequency equal to 0.1 Hz. In 2D simulations a load was applied on the upper face of cartilage fragment corresponding to the value $-4.0397e^{-6}$ N/m in compression, $4.0397e^{-6}$ N/m in tension. For cyclic loading, it was applied at a frequency equivalent to 0.1 Hz

5. Results obtained

5.1. 1D simulations

5.1.1. Compression loads response

Responses from the evidence to compressive loads can be seen in Figure 7. Figure 7a shows the negative shift of the solid component that increases its value negatively with increasing loading time. The time initially is zero (0) for t = 0.67s and reaching figures of $-15x10^{-4}mmN^{-1}s^{-1}$ at the time of maximum loading, t = 45s for x = 0.3. As the liquid flows, the behavior is similar to a linear elastic behavior, because the only component that supports the load is solid.

Figure 7b shows the decrease in pressure due to the fluid outlet presented by the compression of the tissue. For an initial time t = 0.67s the pressure exerted on the fluid in the tissue is 2 MPa, but as the permanence of the compressive, the pressure decreases and takes values close to 0 MPa, thereby external pressure is balanced, also zero.

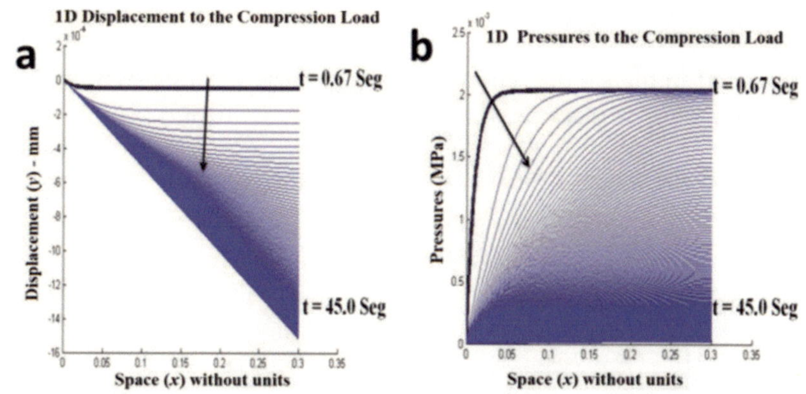

Figure 7. Response of tissue to compression forces. a. Displacement of solid matrix. b. Changes in the fluid pressure in the tissue in presence of the displacement of the same.

5.1.2. Tensil loads response

Figure 8 summarizes the results obtained for tensile strength. Figure 8a shows the positive displacement of the solid component in response to tensile load imposed on tissue. The figure shows how the displacements are positive in presence of tension maintained over time. From an initial value of 0 $mmN^{-1}s^{-1}$ at t = 0.67s, displacement increases up to a maximum of $16x10^{-4}$ $mmN^{-1}s^{-1}$ at a maximum load time t = 45s.

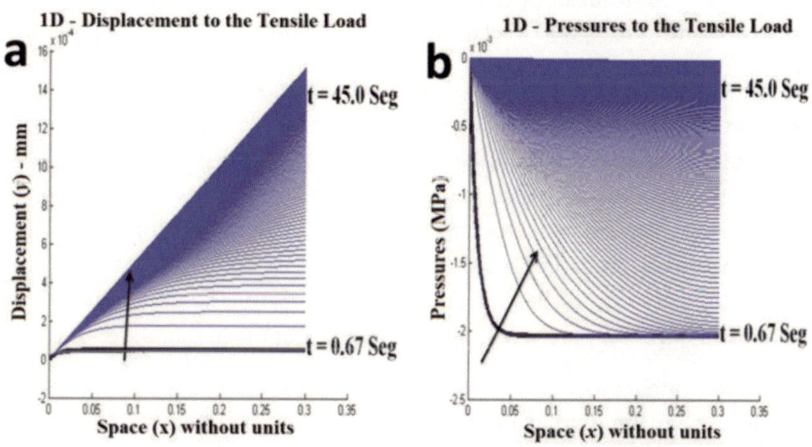

Figure 8. Response of tissue to tensile forces. a. Displacement of solid component tissue upward. b. Change in the fluid pressure generated by the redistribution of fluid in the tissue.

Figure 8b shows the increase of fluid pressure generated by the entry thereof into the tissue due to the pressure difference between the internal and external environment (fluid suction). This pressure difference is caused by displacement of the solid component in the positive direction in response to imposed tensile load. It is shown that this progression of the increase of pressure manifests over time, creating a redistribution of the fluid in the tissue.

5.1.3. Oscillating loads response

Responses obtained in the simulation of oscillating charges can be seen in Figure 9. Figure 9a shows the alternating displacement presented at each oscillation.

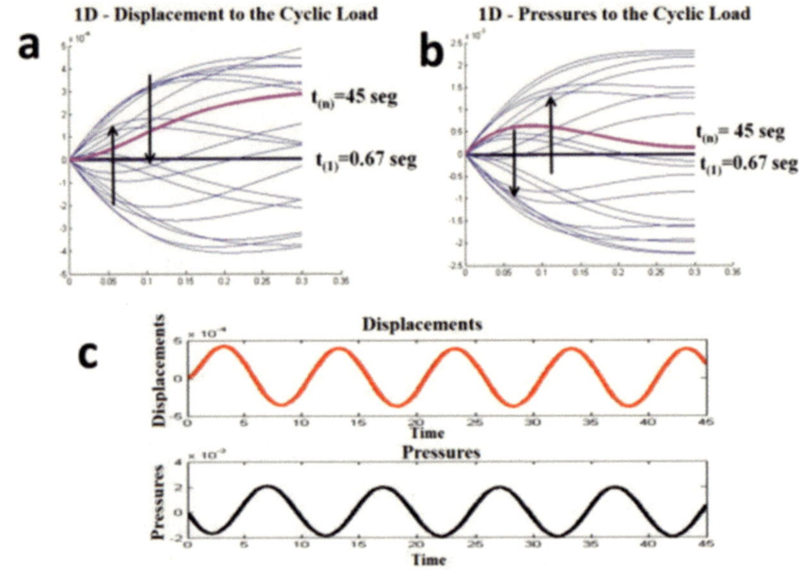

Figure 9. Response to cyclic load. a. Alternate displacement of the solid component. b. Changes in the fluid pressure in oscillate form. c. Delay between the solid deformation and fluid pressure variation.

Initially the movements of the solid are small, close to 0 $mmN^{-1}s^{-1}$ for t = 0.67s. However, they increase over time reaching a value of $6x10^{-4}$ $mmN^{-1}s^{-1}$ or $-6x10^{-4}mmN^{-1}s^{-1}$, at t = 45s, according to the tissue load, tension or compression respectively. Figure 9b shows the development of the process of oscillation in the fluid pressures in response to movement that the solid matrix undergoes upon perceiving the cyclic loading. Initially the pressure lowers and then makes an adjustment that increases, being in inverse phase with the evolution of the deformation of the solid phase of the tissue. I.e., once the solid is deformed in the positive direction, the pressure changes, becoming more negative and vice versa.

The obvious alternating deformation processes of the solid in the face of the application of cyclic loads shows that the displacements are caused by the loads exerted on the tissue and the mobilization of fluid from or into the interior as applied tension or compressive loads respectively. These loads, in turn, generate alternation with respect to each instant of time between the variation of the pressure pattern and the variation of the displacement pattern. However, this action is not in complete phase with displacement. One can appreciate the presence and delay of alternating processes described above, because the equation that represents the displacements is elliptical and corresponds to a displacement equation in space while the equation that represents the pressure corresponds to a parabolic equation and represents a much slower diffusion process than the process of displacement (See Fig. 9c).

5.2. 2D simulations

5.2.1. Compression loads response

Figure 10a corresponds to the displacement of the solid in the *y* axis for each time instant. The displacements are small in the first moments of the load and it can be seen that as the load increases, the time of displacement increases toward more negative values, demonstrating a greater deformation of the solid phase of the tissue. These displacements are produced by the fluid outlet in response to the maintained compression load and it is observed that the greatest displacement occurs in the upper layers of the tissue responsible for receiving the load directly, while the displacement transference is less at a greater tissue depth.

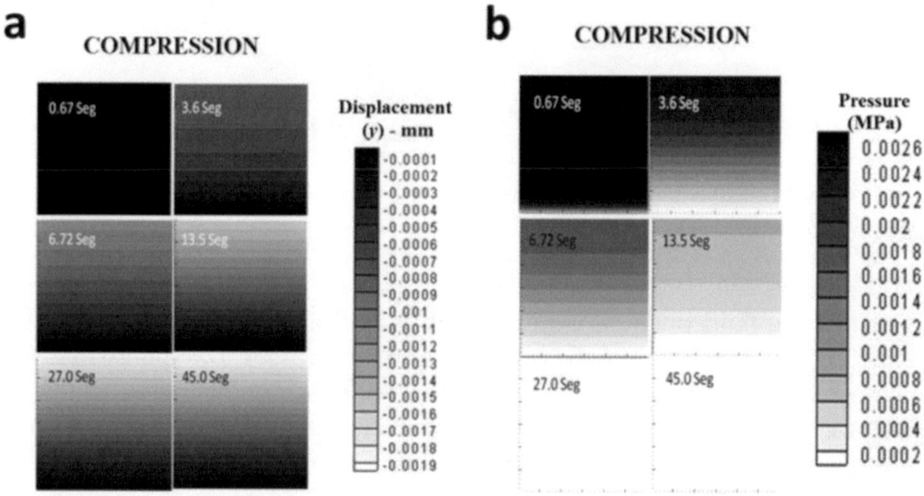

Figure 10. 2D - Response to compressive load. a. *y* displacements during 45 seconds. b. Behavior of the fluid pressure *p* due to its outflow in presence of the load.

Figure 10b shows the behavior of the fluid pressure at each instant of time. It is observed that the pressure decreases rapidly in all tissue layers reaching values close to 0 MPa in a very short time. This variation of fluid pressure corresponds to the decrease of the same in function of the load exerted over the time and poro-elastic tissue behavior which allows the fluid outlet.

5.2.2. Tensil loads response

Results obtained from the tensile simulation are shown in Figure 11. Figure 11a shows the displacement of the solid in the *y* axis for different time instants. There are major shifts in the early stages of loading (t = 0.67s) and displacement decreases with increasing load time. It is further noted that with a sustained tensile load at a given time, the greater displacement or elongation occurs in the upper layers of the tissue where the stress is felt in the first instant, which is why the transfer of deformation is smaller at a greater depth of tissue. In this case the fluid pressures tend to increase because the tissue seeks to balance the inside and the outside environment. Then displacement occurs in response to stress and produces a reorganization of the fluid within the tissue, interfering with the variation in the pressure thereof.

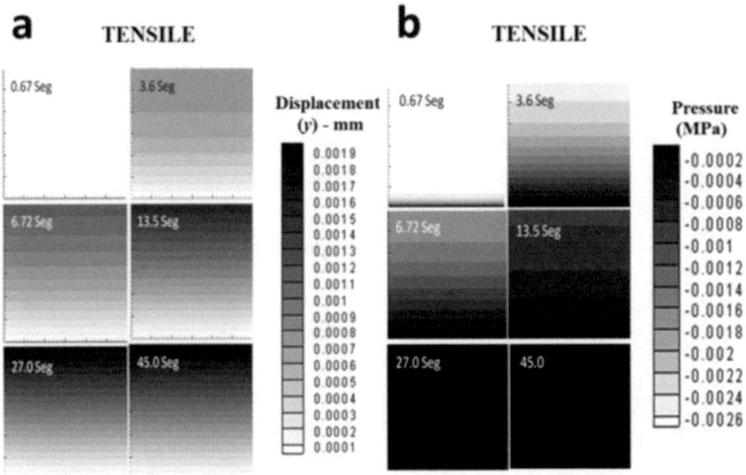

Figure 11. 2D-tensile load response. a. y displacements. b. . Behavior of the fluid pressure p due to its inflow in presence of the load.

Figure 11b shows the behavior of the fluid pressure p for different time instants. It is observed that the fluid pressure is increased by small amounts in response to sustained tensile load. This action is due to the compensation of the tissue in response to deformation in elongation by the solid phase of the tissue that requires a redistribution of fluid into the tissue. However, it is evident that at the end of the tensile load time (approximately at t = 27s), the tissue can't undergo greater deformation in elongation and hence the fluid pressure also tends to stabilize at the inside thereof, causing a steady pressure maintained close to zero which balances with the external pressure.

5.2.3. Oscillating loads response

Figure 12a corresponds to the displacement of the solid in the y axis at different time instants. It is noted that the displacements or deformations occur alternately; at the times 1, 3 and 5 (t = 0.67s, t = 27s = 6.57s respectively) the displacements are made positive, behavior similar to that observed during exposure to stress loads. Conversely, at the times 2, 4 and 6 (t = 3.37s, t = 16.5 s and t = 45s respectively) the displacements are negative, consistent with the behavior exhibited by the matrix to compressive loads.

Figure 12b shows the behavior of the fluid pressure p. Similar to what happened with the deformations of the solid phase, the pressure oscillation in response to cyclic loading imposed on the tissue is evident. Note that at times 1, 3 and 5 (t = 0.67s, t = 6.57syt = 27s respectively) the pressure tends to decrease due to the redistribution of fluid in response to perceived stress loads. Because the loads are not maintained, the tissue can't compensate with fluid inlet from the outside which is why the pressure is not increased. Thus, at times 2, 4 and 6 (t = 3.37s, t = 16.5 s and t = 45s respectively), the pressure tends to increase in response to the compression of the tissue. This is due to the rapidity with which compressive and tensile loads are alternated, loads which prevent the behavior of cartilage during the cyclic loading from exhibiting the behavior that corresponds exclusively to the case of tension or compression.

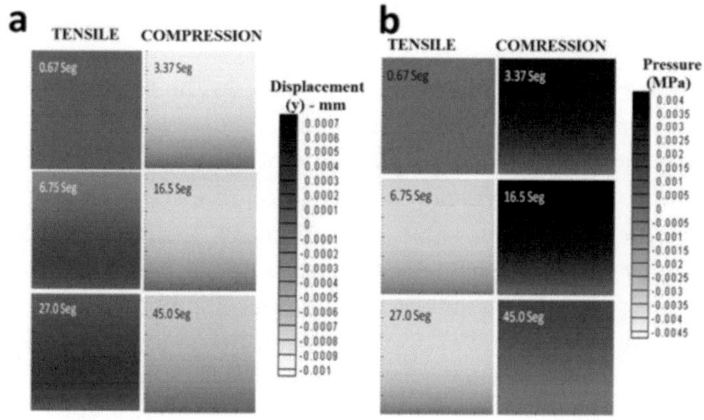

Figure 12. 2D - Response to oscillate load. a. *y* displacements. b. Oscillation of the pressure *p* within the tissue according to the load.

6. Final discussion

There are several theories explaining the behavior of AC in the presence of load conditions, summarized in computational models that include the swelling process and the properties of the anisotropic structure of collagen. The most frequently used tests to determine the mechanical qualities of the AC are the confined compression, the unconfined compression, the indentation and the swelling tests (Wilson, et al., 2005), carried out using numerical approximation tools.

For purposes of meeting the stated objectives, we simulated a condition of confinement of the tissue that allows the flow at the bottom to restrict lateral and bottom movement. Conditions were similar to those reported in practical experiments as the papers presented by Ateshian et al., 1997; Frijns, 2000 and Wu et al., 1999; among others. The data obtained from the simulations confirm the theory of biphasic articular cartilage, first proposed by Mow et al., 1980; and supported by several authors as Haider et al., 2006; Wilson et al., 2005, Haider & Guilak , 2007; Meng et al., 2002; Wu et al., 1997, Terada et al., 1998, Donzelli et al., 1999 and Donzelli & Spilker, 1998; among others.

Results allow us to conclude that articular cartilage exhibits a displacement response of the solid component (matrix) and a variation in the pressure of fluid component due to the exit or entrance thereof, with decreases in pressure in response to compressive loads and increases at the same tensile loads. The displacement is caused by outflow of fluid in response to the maintained compressive load. However it is important to note that once the tissue reaches its maximum displacement, it behaves as a solid rather than as a poro-elastic material. From this point the fluid can't flow out of the tissue because the pressure is balanced with the external fluid begins to bear part of the load.

If this pressure is maintained for prolonged periods dehydration of the tissue may result and cause changes in normal behavior, making it temporarily or permanently more sensitive to injury. These findings support the conclusion that a load consistently maintained for long periods of time or an excessive load that exceeds the characteristics of the tissue once it has reached the maximum possible displacement, can make one more vulnerable to overuse injuries.

The data reported are the beginning of broader work in the study of cartilaginous tissues which can incorporate the cellular component differentially and cartilage's own biochemistry in the model. The results obtained motivate the efforts that currently seek to simulate the production and destruction of the matrix in the presence of mechanical loads, to simulate the restructuring of the same after an injury, to apply mathematical models in the study of cartilage growth and to study their behavior in vitro and in vivo. These lines of research aim to provide a solid foundation for the development of AC experiments *in vivo* and *in vitro* that expands the range of applications of numerical simulation techniques and techniques used in tissue engineering.

Author details

Nancy S. Landínez-Parra
Group of Mathematical Modeling and Numerical Methods GNUM-UN, Faculty of Engineering, National University of Colombia, Colombia

Human Corporal Movement Department, Faculty of Medicine, National University of Colombia, Colombia

Diego A. Garzón-Alvarado and Juan Carlos Vanegas-Acosta
Group of Mathematical Modeling and Numerical Methods GNUM-UN, Faculty of Engineering, National University of Colombia, Colombia

Acknowledgement

The authors wish to thank the Research Division of Bogota (DIB) of the National University of Colombia in the Call for Research for supporting this work under the project "Mathematical Modeling and Simulation of Processes in Mechanical and Biomedical Engineering" and Colciencias 2011 throughout the project "Model for the definition of term of the Layout of a manufacturing cell through set theory and optimization" who contributed to the financing the chapter.

7. References

Almeida, E.S. & Spilker, R.L. (1998). Finite element formulations for hyperelasticbiphasic soft tissues transversely isotropic. *Computer methods in applied mechanics and engineering.* 151, 513-538

Andriacchi, T.P., Mûndermann, A., Smith, R.L., Alexander, E.J., Dyrby, C.O. & Koo, S. (2004). A framework for the in vivo pathomechanics of osteoarthritis at the knee. *Annals of Biomedical Engineering*. 32, 447–457.

Ateshian, G.A., W.H. Warden, J.J. Kim, R.P. Grelsamer & V.C. Mow. (1997). Finite deformation biphasic material propertieso f bovine articular cartilage from confined compression experiments. *J. Bimechanics*. Vol 30. Nos. 11/ 12. pp 1157-1164. Published by Elsevier Science

Boschetti, F., Miotti, C., Massi, F., Colombo, M., Quaglini, V., Peretti, G.M. & Pietrabissa, R. (2002). An Experimental Study on Human Articular Cartilage Permeability. *Proceedings of the Second Joint EMBS/BMES Conference*. Houston, TX, USA. October 2002. 23-26

Carter, D.R. & Wong, M. Modelling Cartilage Mechanobiology. (2003). *The Royal society. Philos Trans R Soc Lond B Biol Sci.* 2003 September 29; 358(1437): 1461–1471.

Chan, B., Donzelli, P.S. & Spilker, R.L. (2000). *A Mixed-Penalty Biphasic Finite Element Formulation Incorporating Viscous Fluids and Material Interfaces. Annals of Biomedical Engineering,* Vol. 28, pp. 589–597

Doblaré Castellano, M. (*3 de noviembre del año 2005*). Sobre el Modelado en Biomecánica y Mecano-biología. *Discurso de ingreso a la Real Academia de Ciencias Exactas, Físicas, Químicas y Naturales de Zaragoza.*
http://www.unizar.es/acz/02AcademicosNumerarios/Discursos/Doblare.pdf

Donzelli, P.S., Spilker, R.L., Ateshian, G.A. & Mow, V.C. (1999) Contact analysis of biphasic transversely isotropic cartilage layers and correlations with tissue failure. *Journal of Biomechanics*. 32, 1037-1047.

Donzelli, P.S. & Spilker, R.L. (1998). A contact finite element formulation for biological soft hydrated tissues. *Computer methods in applied mechanics and engineering*. 153, 63-79

Frijns, A.J.H. (2000). A Four-Component Mixture Theory Applied to Cartilaginous Tissues. Tesis Doctoral. *Eindhoven University of Technology.*

Garzón, D.A. (Mayo de 2007). Simulación de Procesos de Reacción-Difusión: Aplicación a la Morfogénesis del Tejido Óseo. Tesis doctoral. *Centro Politécnico Superior de la Universidad de Zaragoza*. Zaragoza.

Guilak, F., Jones, W., Ting-beall, P. & Lee, G. (1999). The deformation behavior and mechanical properties of condrocitos in articular cartilage. *Osteoarthritis and Cartilage*. 7, 1: 59–70. January 1999

Haider, M.A. & Schugart, R.C. (2006). A numerical method for the continuous spectrum biphasic poroviscoelastic model of articular cartilage. *Journal of Biomechanics*. 39, 177–183.

Haider, M. A. & Guilak, F. (2007). Application of a three-dimensional poro-elastic BEM to modelling the biphasic mechanics of cell–matrix interactions in articular cartilage. *Computer methods in applied mechanics and engineering*. 196, 2999–3010.

Koenig, C. (Noviembre de 2011). *Curso de Histología*. Oficina de Educación Médica. Escuela de Medicina. Pontificia Universidad Católica de Chile. Consultado el 25 Noviembre 2011. http//escuela.med.puc.cl

Lawrence, R.C., Helmick, Ch.G. & Arnett, F.C. (1998). Estimates of the prevalence of the arthritis and selected musculoskeletal disorders in the United States. *Arthritis Rheum;* 41: 2213-2218.

Levenston, M.E., Frank, E.H. & Grodzinsky, A.J. (1998). Variationally derived 3-field finite element formulations for quasistaticporo-elastic analysis of hydrated biological tissues. *Computer methods in applied mechanics and engineering*. 156, 231-246.

Martín-Hernández, C. (2002). Estudio Mecánico, Histológico, e Histo-morfométrico del Regenerado de Cartílago a Partir de Injertos de Periostio Invertido. Tesis Doctoral. *Universidad Autónoma de Barcelona.*

Meng, X.N., Leroux, M.A., Laursen, T.A. & Setton, L.A. (2002). A nonlinear finite element formulation for axisymmetric torsion of biphasic materials. *International Journal of Solids and Structures*. 39, 879–895.

Meyer, U. & Wiesmann, H.P. (April 11, 2006). Bone and Cartilage Engineering. (1 edition) ED Springer. ISBN 978-3-540-25347-1. Germany. pp 25-27.

Mow, V.C., Kuei, S.C., Lai, W.M. & Armstrong, C.G. (2006) Biphasic creep and stress relaxation of articular cartilage in compression: theory and experiments. *Journal of Biomechanical Engineering*. 102: 73-84. X

Nordin M. & Frankel, V. (2004). Biomecánica Básica del sistema Musculoesqueletico. *ED McGrawn-Hill Interamericana.* pp 61.

Radi, M. (1998). *Three-Dimensional Simulation of Thermal Oxidation*. Dissertation. Institute for Microelectronics. Faculty of Electrical Engineering and information technology 1998. Consultado el 25 de Abril de 2012. http://www.iue.tuwien.ac.at/phd/radi

Sopena-Juncosa, J.J., Carrillo-Poveda J.M., Rubio-Zaragoza M., Redondo-García J. I., Serra-Aguado I. & Soleri-Canet I. (2000). Estructura y función del cartílago articular. *Portada: Armas Frente a la Patología Articular.*

Terada, K., Ito, T. & Kikuchi, N. (1998). Characterization of the mechanical behaviors of solid-fluid mixture by the homogenization method. *Computer methods in applied mechanics and engineering*. 153, 223-257.

Van der Meulen, M. & Huiskes, R. (2002). Why Mecanobiology? A survey article. *Journal of Biomechanics*. Volume 35, Issue 4, Pag: 401–414.

Wilson, W., Van Donkelaar, C.C., Van Rietbergen, B., Itoa, K. & Huiskes, R. (2004). Stresses in the local collagen network of articular cartilage: a poroviscoelastic fibril-reinforced finite element study. *Journal of Biomechanics*. 37, 357–366.

Wilson, W., Van Donkelaar, C.C., Van Rietbergen & Huiskes, R. (2005). A fibril-reinforced poroviscoelastic swelling model for articular cartilage. *Journal of Biomechanics*. 38, 1195-1204.

Wilson, W. (2005). An explanation for the onset of mechanically induced cartilage damage. Tesis Doctoral. *Technische Universite it Eindhoven.* Eindhoven.

Wu, J.Z., Herzog, W. & Epstein, M. (1997). An improved solution for the contact of two biphasic cartilage layers. *Journal of Biomechanics.* Vol. 30, No. 4, pp 371.-375

Wu, J.Z., Herzog, W. & Epstein, M. (1999). Modelling of location- and time-dependent deformation of chondrocytes during cartilage loading. *Journal of Biomechanics.* 32: 563-572.

Wu J.Z. & Herzog, W. (2000). Finite Element Simulation of Location and Time-Dependent Mechanical Behavior of Chondrocytes in Unconfined Compression Tests. *Annals of Biomedical Engineering*. Vol. 28, pp. 318–330

Wu J.P. & Kirk, T.B. (November 2001). A Study of the Shape Change of the Sheep Chondrocytes with Application of Compression to Cartilage. *Seventh Australian and New Zealand Intelligent Information Systems Conference,* 18-21. Perth, Western Australia